Technologies of Control

Studies in Communication and Society

A series edited by

Vincent Mosco and Janet Wasko

Communication and Latin American Society: Trends in Critical Research, 1960–1985
Edited by Rita Atwood and Emile G. McAnany

The Political Economy of Information
Edited by Vincent Mosco and Janet Wasko

Technologies of Control: The New Interactive Media for the Home
Kevin G. Wilson

Kevin G. Wilson

Technologies of Control

The New Interactive Media for the Home

The University of Wisconsin Press

Published 1988

The University of Wisconsin Press
114 North Murray Street
Madison, Wisconsin 53715

The University of Wisconsin Press, Ltd.
1 Gower Street
London WC1E 6HA, England

First printing

Printed in the United States of America

For LC CIP information see p. 180

ISBN 0-299-11370-1 cloth; 0-299-11374-4 paper

TO CARMELA

Contents

Preface

This book was conceived as an alternative to mainstream research into the social impact of home networking. The sheer quantity of printed material on home networking is witness to wide appreciation in North American society of the need to understand these systems. A review of this literature reveals, however, that such an awareness has not been translated into an adequate understanding of the importance of these systems to society. Home networking is perceived as a simple extension of contemporary mass media and telecommunications technology, albeit in the form of an innovative hybrid. It follows from this premise that a public policy for home networking may be adopted in an ad hoc, wait-and-see, piecemeal fashion, utilizing existing conceptual frameworks and policy initiatives to ensure that this capability will be harnessed for the benefit of society. To my way of thinking, this approach represents a fundamental misunderstanding of the nature of home networking and underestimates the challenge it presents to society.

The following analysis attempts to fill a gap in critical research on home networking technology by discussing the importance of these systems to practices of social control. Although this issue has been examined in the literature on home networking, it has not been subjected to systematic analysis. My work documents structural trends in the organization of an emerging interactive services industry which suggest that information generated by these systems will be applied to practices of social control. This analysis is combined with a critical reappraisal of the modalities of social control and computer networking.

The present study demonstrates that a discussion of social control and computer networking need not and should not be limited to the terms of the privacy debate. This debate posits an understanding of data-based social control solely along an axis relating the individual to any number of bureaucratic institutions. It is unable to account for practices of aggregate social control which do not rely on information retained in personally identifiable form. A whole gamut of social-control activities supported by the computer, here called practices of *social management,* are excluded from public scrutiny. This study accepts the validity of the privacy debate as a partial framework for understanding social control in home networking, but argues that home networking will feed information into a number of computer-based activities of aggregate social simulation, social and technological forecasting, and long-term social forecasting and planning, which must also be considered practices of social control.

In order to substantiate this argument, materials are brought to bear from related contexts in the larger debate on computers and society. These include frameworks drawn from the critique of management science and computerization concerning social control and the labor process, as well as those of systems analysis and organizational complexity. Attempts in political economy to understand the role of computer technology in the growth of corporate-sector institutions are introduced as a way of explaining, at a macrosocietal level, the vital contribution that computer systems have made to the maintenance of the cybernetic economy of late capitalism. The critical and innovative thrust of this book resides in my integration of these materials to reformulate, in a more comprehensive fashion, the social control implications of computer networking for the home. To the extent that I document the threat posed by interactive systems to personal privacy, this study may also be viewed as a contribution to privacy policy. For those researchers and citizens concerned with this issue, the evidence I present here suggests that new initiatives will be required if the privacy interest is to be defended in home networking.

Acknowledgments

I would like to acknowledge the contributions of three individuals who, in different ways and at different stages in my research, made significant contributions to its progress and success: Abbe Mowshowitz, Marike Finlay, and Irena Bellert. I consider myself to have been extremely fortunate to have worked with each of them and sincerely hope that our personal and professional contacts will be maintained in the years to come.

Thanks are also due to the Graduate Program in Communications, the Graduate Faculty, and the Fellowships Office at McGill University, the Max Bell Foundation, and as well as to Télé-université for assistance in editing the manuscript.

Finally, I would like to acknowledge the contribution of my wife, Carmela, to whom I dedicate this book. Without her love and support it could not have been written.

Technologies of Control

Introduction

A consensus on interactive technology is forming and solidifying under the auspices of technological forecasting efforts, technology assessment studies, and policy research. This assessment is unique in that it is occurring several years, if not decades, in advance of the introduction of interactive technology into the home. A recent Institute for the Future study (Tydeman et al. 1982, 287–300) on teletext and videotex included a bibliography of over three hundred titles, the vast majority of which were devoted specifically to interactive technology. The discourse on interactive systems and the home incorporates forecasts and likely scenarios for future implementation of this technology, as well as policy evaluations of its social impact. The extent to which home networking has been assessed, analyzed, and written about in anticipation of its future introduction has few, if any, parallels among modern technologies, communicational or otherwise.

In spite of their numbers, these studies circumscribe a much larger field of potential analysis in order to avoid critical reflection on interactive systems. Research has been biased by the implicit assumption that technology is an autonomous force with the power to achieve a unilateral impact upon society. Such a causal, unilinear perception of the society/technology interface has impoverished the social analysis of new communications technology (and, in particular, the social analysis of home networking) by idolizing technology and obscuring the economic and political content of these systems. The actual powers, values, and interests which have combined to develop interactive technology are obscured in technological forecasting, technology assessment, and policy research. Such studies do not analyze the social significance of home networking in a comprehensive, critical fashion, but, instead, consider interactive systems in an ahistorical, decontextualized manner conducive to accelerated technology transfer and development.

Bias is also apparent in the pervasive future orientation of research on home networking. Studies neither identify nor explore the factors which demand, or even compel, the realization of these systems under specific economic, political, and social conditions as a way of interpreting their meaning for society. This historical context of the implementation of technology is ignored in research preoccupied with predicting and assessing future trends and impacts. The *futures* paradigm of research does not inquire into the factors which foster and motivate social investment in technology.[1] Moreover, it does not acknowledge the contingent relationship between technological innovation and the social forces which impel it.

Although there exist numerous examples of the futures approach to home networking, a recent (1979) Canada Science Council report entitled *A Scenario for the Implementation of Interactive Computer-Communication Systems in the Home* particularly exemplifies this kind of research. (see Wilson 1982, 116–75). In this report, the whole question of the social impact of home networking, not to mention that of its social significance, is framed by the technological imperative of orderly technology transfer. It is assumed that a positive impact can be assured through rational planning of the implementation of home networking technology. There is no broad consideration of the social significance of home networking. What marks this report as typical of research in the field is its preoccupation with the means toward achieving technology transfer, while discouraging the analysis of what the goals of this process should be. Its stated objectives are the reduction of uncertainty, the elimination of ambiguity, and the orderly transfer of technology. For example, the report states the following:

> The purpose of this paper . . . is primarily to demonstrate that implementation *can and must* [emphasis added] take place in an orderly way. (8)

> The intent is to demonstrate that the *ambiguity* [emphasis added] of the present situation can be diminished—if not eliminated—when a policy for implementation is adopted. (8)

> As noted in the previous paper, the framework of regulation is *confused* and the development of a *coherent* [emphasis added] policy will be very difficult. (9)

In the absence of any explicit commitment to a discussion of the social "ends" of technology, what would normally be considered "means"-related

1. The term *futures research* is used here to designate the fusing of technological forecasting, policy discourse, and sociological discourse (the theory of postindustrial society) which occurred in the mid-1960s under the auspices of a new discipline, futures research. Elsewhere I have analyzed this discipline as a discourse, demonstrating how it has restricted public debate on new technologies. The discourse purports to assess the social implications of technology in an objective manner, while in fact it performs ideological labor for its corporate sponsors by creating a social consensus on new technology conducive to early implementation. See Kevin G. Wilson, "Futures Research and Interactive Videotex: An Analysis of the Discourse on New Communications Technology and Society," unpublished paper, 1982.

issues become "ends" in themselves. This is true to such an extent that the Science Council report elevates the "orderly" transfer of technology to the status of a social goal. Consequently, the diminishment of ambiguity, the elimination of confusion, and the planning for an orderly transfer of technology become ends in themselves. Superseding any discussion of ultimate social goals are references to the attainment of the "*maximum* benefit" of technology (p. 37), reinforced by a textual commitment to what will be "*optimal* for our needs" [emphasis added] (6). What constitutes "maximum benefit" and the "optimal" in terms of needs is never discussed. Elaboration on these issues would have constituted an inexpedient digression into a debate on social goals inconsistent with the report's major aims (i.e., the justification and promotion of technological advance), a conviction which seems to predominate in most of the literature in the field. Indeed, the emphasis on optimization and maximization betrays the text's affinity for the language, if not the priorities, of the corporate sector.

This balance-sheet model of the relationship between technology and society does not even attempt to obscure the report's advocacy function. Moreover, neither the benefits nor the disadvantages of technology transfer are identified. Instead, the report transforms and reduces the social implications of videotex into a series of recommendations designed to ensure implementation. Although the report is not prepared to catalog social costs, it is prepared to inscribe them semantically in the text as the antithetical counterpart to an orderly transfer. The Science Council report formulates its own structure of the social implications of technology implementation: it associates an orderly transfer of technology with a maximization of benefits. However, it merely assumes that an orderly transfer of technology will be synonymous with the realization of social needs, without demonstrating how or why this should be the case.

In specific terms, the Science Council report—like other studies—obscures the fact that technology implementation may constitute a response to conflicts and problems in the social order. These problems may be studied at various levels (e.g., at a micro level of competing industries or interest groups, or at a macro level of ongoing crisis in the economic order). As Innis (1951) has demonstrated, the introduction of new technologies is often characterized by institutional conflict and the rise to power of previously marginal institutions. The implementation of technology may threaten the hegemony of established groups or institutions. Conversely, the same technology, once perceived as a threat, may be harnessed by institutions as a response to threats to their hegemony.

The futures paradigm of research produces a limited and biased perception of home networking by suggesting that interactive systems will affect society unilaterally and that the only problem is to ensure orderly transfer. A more critical approach to the study of the society/technology interface would be to replace an analysis of social impacts with an examination of the social significance of technology. The social significance of emerging technology is comprehended when technology is interpreted in the context of its sociohistorical development. The principal task of critically informed research into home networking must be to relate this technology to the economic and political contexts of its development and thereby decipher its social content as something which is already manifest and susceptible to analysis.

As one prominent study (Tydeman et al. 1982, 3) has noted, teletext and videotex are not "new" technologies. The technological basis for interactive systems can be traced back over twenty years to the development of time-sharing on computers. The idea of a large-volume, automated information retrieval system not unlike videotex is often traced to Bush (1945). Although interactive technology has not yet been established in the home, efforts to promote and ensure such a development have been a matter of public record since the mid-1960s, when computer enthusiasts first suggested the possibility of a "computer utility" (McCarthy 1968, 220–24).

In recent years, the development of teletext and videotex graphic protocols has become a national priority of governments in Canada, Britain, and France. Private-sector interest on the part of communication carriers, publishers, advertisers, retailers, banks, and manufacturers has become more intense with each year. The process of research and investment into home networking is sufficiently advanced to justify analysis. Whereas futures research (incorporating technology assessment, technological forecasting, and some policy research) proposes to evaluate the social future of interactive technology, the present study examines its social significance as something already manifest in the research, decision-making, and investment processes promoting it.

Williams' (1974) analysis of the social history of television in the United States and Britain is a pertinent example of critical research into the technology/society interface. By rejecting a "determinist" view of technological development, as well as what he terms the "symptomatic" view, Williams restores "intention" to the study of technological development. Moreover, he identifies this factor with the earliest stages of research on and investment in technology, stating his position in the following terms:

> Such an interpretation would differ from technological determinism in that it would restore *intention* to the process of research and development. The technology would be seen, that is to say, as being looked for and developed with certain purposes and practices already in mind. At the same time the interpretation would differ from symptomatic technology in that these purposes and practices would be seen as *direct*: as known social needs, purposes and practices to which the technology is not marginal but central. (14)

The task for an analysis of home networking must be to identify those intentions which are latent in decisions to sustain and advance the development of interactive technology. In the case of videotex and teletext technologies, the process of technological development is ongoing. This contrasts somewhat with the case Williams studied, where television already existed as an established institution. The fact that home networking is still an evolving technology does not present any major problems to this type of analysis, since it has been "evolving" for over twenty years.

Williams has made an important contribution in another respect by signaling the relationship between technology and crisis in the social order. He demonstrates that the development of television was both "a response to the development of an extended social, economic and political system and a response to a crisis within that system" (21). This crisis, he maintains, involved the integration of the privatized home into the political and ideological life of the nation-state.

A number of parallels may be drawn between Williams' analysis of television and the case of home networking. Although separated in time, both communication technologies are targeted for consumption in the home. Both technologies, as Williams demonstrates so well in the case of television, represent powerful instruments of social control. Moreover, at a macro level of analysis, Williams establishes the connection between crisis in the social, economic, and political system and the value of television as an institution of social control capable of defusing it.

Williams' work establishes an important objective for research into interactive technology: it must explore the relationship between interactive systems, social control, and crisis in the social order. Preliminary research suggests that home networking has been identified with issues of data-based social control. For the most part, however, this concern remains a rhetorical one, unsubjected to detailed analysis. Parkhill's (1979) article may be considered as typical of this pattern in the literature on home networking. Parkhill raises the issue of electronic social control with references to "big brother" and Orwell's *1984*—allusions which are fairly common in other studies as well. He states:

> There is an uncomfortable aura of 1984 in all this. Thus it is not difficult to visualize how the computer utility might be perverted until it degenerated into nothing more than an instrument of total political control—the omnipresent eyes and ears of "big brother." (89)

At least two features of Parkhill's commentary mark it as representative of the manner in which the relationship between social control and home networking has usually been analyzed. First, the modalities of social control are identified and illustrated at a merely superficial level through references to literature. In effect, there is no analysis; it is replaced with literary imagery. Second, the interpretation given to this imagery casts the problem of electronic social control in terms of a threat solely from the public sector, the state. In so doing, Parkhill obfuscates an equally coercive site of control: the bureaucratic structures of private-sector corporations. Moreover, this interpretation ignores the symbiotic relationship (in advanced industrial states) between the state and these private-sector corporations, whose interests frequently converge on matters concerning the formation and monitoring of public opinion. Although acknowledging the issue of social control, Parkhill minimizes it by representing it in a limited context of state totalitarianism. One may assume that totalitarianism is to be identified with communism, which explains the reference to a "perverted" development of the computer utility. Parkhill implies that social control is a matter concerning the citizen and the state which may be dealt with by private-sector ownership and operation of interactive systems. Accordingly, the major safeguard required to assure a democratic society is one that restricts information transfers between public and private sectors.

Although Parkhill's remarks are but a single register of public concern with social control in home networking, such a concern is far from unique to his work. The issue is fairly prevalent in most of the literature which purports to analyze interactive systems in a social context, receiving its most concrete formulation in the privacy debate.

A notable exception to this trend in research is Mosco's 1982 book, *Pushbutton Fantasies*. Unlike the dominant futures paradigm of research, Mosco's work analyzes interactive systems in the context of a critical theory of society. Mosco reviews the theoretical underpinnings of mainstream mass society theory, contrasts this with radical approaches to social theory, and then demonstrates how and why these approaches lead to different interpretations of what we would term the "social significance" of home networking.

Mosco then analyzes interactive systems in three relevant contexts: the home, the workplace, and international economic development. His discussion of interactive systems in the home, the most pertinent to my study, contrasts a *mass society* view of the media with the *audience commodity* perspective elaborated by Dallas Smythe as a way of suggesting a third alternative based on the concept of the *audience sphere*.

The social control issue could be formulated in terms of a social dynamics of public versus private spheres. For example, considerations of privacy and home networking are clearly grounded in the threat posed by these systems to the private sphere as more and more personal information becomes subject to bureaucratic consumption in private- and public-sector institutions. Mosco notes that the historical trend in the age of mass media has been toward a decline of the public sphere because of economic pressures to deliver audiences to advertisers: "The central *public* debates in capitalism are media events in which candidates for public office are judged on their ability to deliver, not on a policy promise, but an audience to advertisers" (111).

Similarly, the private sphere is subjected to economic pressures as interpersonal relations are increasingly mediated through commodities. The key to understanding this convergence, which is also a diminishing of public and private spheres, is the concept of the audience sphere. Mosco identifies this sphere with a generalized pressure exerted in capitalist economies to transform all human activity into marketable commodities:

> To suggest that the audience sphere invades both public and private domains preserves the meaning of the terms and avoids confusing implications about the relationship between public and private spheres. The "fall of public man" results not from the intrusion of personalities into politics, but from the intrusion of the audience commodity. It is not intimacy that overwhelms the public sphere, but a false intimacy that builds the audience commodity. Likewise, it is not the intrusion of the public sphere that destroys intimacy, but a false sense of community that creates neither public spirit nor private life, but the audience commodity. (111)

The analysis of the social significance of home networking is illuminated by the concept of the audience sphere and its relationship to public and private spheres of human activity. It is clear that interactive systems will be subject to similar economic pressures designed to transform human activities into marketable commodities. It remains to be determined, however, precisely how these pressures will manifest themselves in the context of home networking.

As a mass media scholar, Mosco correctly identifies the role that interactive systems will play in delivering very specialized audiences to advertisers for particular products and ad campaigns as a way of refining the audience production process. In the case of political polling through interactive systems, he notes that an illusion may be created of democratic participation—when, in fact, these systems will institutionalize a privatization of the political. A further development will be a diminishing of the distinction between commercials and programming with the advent of the "infomercial" (105). From my perspective, Mosco has also correctly identified the value of the return channel of interactive systems through such concepts as audience production, political polling, and changes in the character of programming. Moreover, he has situated these developments in their macrosocietal context of corporate capitalism. In order to consider the social significance of home networking, I have elected to pursue the question of the two-way channel (i.e., the return flow of information) in contexts other than that of audience production (the primary focus of Mosco's analysis).

Although information generated by interactive systems has an obvious value to advertisers desiring to reach targeted markets, this is not the only potential use of such information by the corporate sector, particularly in the case of interactive systems which do not piggyback on existing television technology. (For example, current marketing trends suggest that home networking will develop first out of an existing population of microcomputer users who will purchase modems for communication purposes.) Two considerations from the literature on computers and society suggested ways of pursuing and contextualizing the significance of the return channel of interactive systems: (1) the interrelationship of computers and systems of *public surveillance* and (2) the role of computers in social management. The first of these concerns the issue of personal privacy in computerized data banks as elaborated since the 1960s, and the second refers to the role played by the computer in social simulation, forecasting, and planning (i.e., the critique of systems analysis). Because information generated by interactive systems will be channeled into systems of public surveillance as well as into social research for social management, the analysis of home networking must take these practices into account. Here I have considered both the practical question of how interactive systems will furnish vital information to these practices and the critical theoretical question of how a critique of public surveillance and social management may enhance an understanding of the social significance of home networking.

In chapter 1, major trends in the videotex industry are examined, incorporating an analysis of developments in technology, markets, industry competition, ownership, and investment. A scenario for the structural organization of a future videotex industry is proposed, based on an analysis of these trends. Instead of elaborating on the potential of new technology, I have considered those factors which may constrain or limit its implementation. The lack of a ready mass market for interactive systems will continue to slow the technology transfer process. But this factor will not forestall the advance of home networking. Problems in the marketing of interactive systems will, instead, ensure that only the very largest corporations in the fields of electronics manufacturing, mass media, and communications will participate.

Moreover, a tremendous amount of transindustry cooperation will be required. Given the difficulty of financing a videotex industry in the absence of a ready mass market, the return flow of consumer preference information and polling data will be crucial to the success of home networking. This return flow of information is termed the *informational commodity* of interactive systems as a way of suggesting its value to the videotex industry. When resold to other firms or processed internally (as for marketing purposes, long-term forecasting and planning), the informational commodity will be a source of added value to the industry. It will be very difficult, if not impossible, to constrain its use. Once this information is stored in computer data banks, it will serve as input to computer-based systems of public surveillance and social management. These activities will significantly enhance the corporate sector's capacity to manage the social order.

An obvious public concern—and the one which has received the most attention in policy discussions of the return flow of information from home networking—is privacy protection. In chapter 2, the North American privacy tradition is reviewed as a basis for understanding what constraints, if any, will be placed on corporate use of the informational commodity. This critique demonstrates the historical failure of privacy policy to guarantee its stated objectives, with ominous implications for the protection of personal privacy in home networking.

In chapter 3, current trends in creating privacy policy for home networking are examined. James Rule's concept of public surveillance, based on the problem of agency control of mass clienteles, is the theoretical backdrop for the discussion of social control and privacy protection in this chapter. A significant development will be the transvaluation of the privacy interest into an economic one: consumers will be encouraged to sell off personal information for economic benefit as a way of circumventing the privacy issue. Given this trend, together with the historical failure of the North American privacy policy, individuals may find it increasingly difficult to control information about themselves as they seek participation in home networking. In this chapter, the problem of data-based social control is framed by the issue of privacy protection. The conceptual underpinnings for this view of electronic social control are grounded in a liberal tradition of respect for the autonomy and freedom of the individual.

In chapter 4, an alternative conceptualization of social control and home networking is proposed, based on the use of information generated by these systems for corporate-sector social management. Social management designates uses of the informational commodity for social simulation and long-term social forecasting and planning. Although these activities do not threaten individual liberties in a direct fashion, they nevertheless constitute practices of social control which threaten the balance of power between corporate and noncorporate sectors. The concentration of these informational resources in the corporate sector will foster monopolies of knowledge antithetical to the principles of democratic decision making and participation historically valued as institutions of North American society.

The development of home networking will be situated in a macrosocietal context of the information economy of cybernetic competition. The incorporation of the home into computer networks is consistent with trends in the evolution of the communications infrastructure, geared to the demands of the cybernetic economy of late capitalism. To the extent that home networking generates much information which will serve as input to macrosocietal, long-term social management, home networking represents a partial response to an ongoing crisis in the economic order of late capitalism.

Finally, chapter 5 identifies a number of general trends in communications which will have an impact on home networking. These include (1) a politics of deregulation, (2) the commodification of communications, (3) the fragmentation of the political sphere and the crisis of pluralism, and (4) efforts to ensure property rights to electronic information. The repercussions of these trends and their implications for home networking are considered.

To summarize, the approach represented in this book is one informed by the work of Williams (1961, 1974) and, to a lesser degree, that of Innis (1951). Both theorists concur on the importance of understanding technological innovation in a context of both limitations and crisis. New technologies may provoke as well as respond to crisis. Whereas Williams' work has emphasized the notion of intention in technology development, Innis' has stressed conflict and the introduction of technology. The analysis of emerging interactive technology must take into account crisis, conflict, and intention in the development of such technology. Such an analysis distinguishes the following study of the social significance of home networking from the future-oriented assessment of social impacts.

1

The Videotex Industry

Future Trends and Market Structure

1.1 Commercial Interest in Home Networking

In recent years, the development of Prestel in Britain, Teletel in France, and Telidon in Canada (all heavily subsidized by their respective governments) has made researchers focus on the communications networks of the future. In essence, these systems represent different technological solutions to the problem of encoding (and decoding) graphic information into digital form for transmission by electronic communications media. The graphic protocols that have been developed are versatile because they are compatible with a variety of communication networks: broadcast (TV, MDS, DBS, FM radio); cable television; switched telephone networks; and packet switched networks (Tydeman et al. 1982).

The importance of this technology is primarily a function of its graphics capability. Proponents of the technology recognized immediately that graphic protocols made this an appealing domestic technology. In retrospect, it appears that engineers were naive and overly optimistic in their belief that graphic images alone would be sufficient to bring computer networks into the home. Subsequent research has revealed that marketing computer-based services and information retrieval will be more difficult than offering low-cost decoders for sale to consumers. Minor problems such as the consumer's ignorance of his need for these services will delay home networking for the immediate future. Studies have suggested that certain services, such as teleshopping, will require a higher quality, "photographic" resolution of images than is currently available with videotex and teletext (Ash and Quelch 1982, 85). In addition, the limitations of current channels into the home (paired wire and coaxial cable) may restrict or forestall the success of home networking, at least in the short term. It may be that the full range of services which have been foreseen in connection with home networking will be contingent upon optical fiber technology. Nevertheless, the graphic capabilities developed for videotex and teletext are significant: not only have they renewed interest and spurred investment in a very general way, but, in particular terms, they have justified an initial stage of market testing into home networking.

The object of industrial and governmental strategy is to integrate the home into an infrastructure of computer-based communications. The vehicle is graphics. The perception has been that graphic presentation and display are essential to acceptance by consumers of networking services. It is, therefore, in the context of innovation and national promotion of graphic protocol technology (Telidon, Prestel, and Teletel) that trial testing of interactive networking has begun. This technology marks the current stage in a project which has been foreseen in relatively concrete terms since the 1960s (i.e., time-sharing and the computer utility concept).

Twenty years ago, the literature on home networking focused on the computer utility. Today the favored topics are videotex and teletext. The shift from the discussion of an abstract concept, the computer utility, to the discussion of specific technological capabilities, videotex and teletext, is itself suggestive of the progress which has been made toward eventual home networking. Both videotex and teletext are computer-based, delivering information to the home in digital form. The technologies differ, however, in one important respect: teletext is unidirectional and fundamentally a broadcast technology; videotex is two-way or bidirectional, implying that one can "talk back" to the broadcasting center or host computer.

Although videotex will be some ten or fifteen years behind teletext in its implementation, it is considerably more significant because it will operate in the so-called interactive mode. This means that the home will soon be integrated into the computer networks of the future. Digital message networks have existed for some time in government and business communications systems, but the merging of the home into such networks vastly increases the computerization of society.

Home networking is here viewed as the integration of the home into an infrastructure of corporate-sector, computer-based communications. Factors such as technology, potential services, markets, and ownership patterns are analyzed in this chapter, with the goal of describing emerging trends in the videotex industry and ultimately identifying the relationship between social control and the informational commodity of interactive systems for the home.

1.2 Home Networking and Interactive Systems

The technologies which comprise home networking are divisible into two generic classes: interactive videotex and interactive media systems. The difference between them lies in the character of the information displayed on home screens. Videotex information is either textual (lines of text) or graphic (pages of graphics encoded and decoded through available graphic protocol technology) and is transmitted to the home in digital form. Interactive media technology transmits analog television images to the home. In both cases, messages may be transmitted back to a host computer or a broadcast center in digital form, making this technology interactive. A well-known example of an interactive media system is the QUBE trial in Columbus, Ohio. This system was interactive, but did not incorporate the display of graphic or textual information.

These generic differences are significant for research because they suggest an appeal to different communicational rituals (such as watching TV, reading a newspaper, or playing a video game) and involve different industries (e.g., the broadcast industry as opposed to the newspaper industry). Research has tended to either treat videotex and teletext as a single technology and interactive media as another or to mix these technologies and study them as essentially the same phenomenon, without acknowledging important distinctions. Because teletext does not incorporate an interactive return capability, it is not relevant to a discussion of home networking. On the other hand, the study of home networking must include the study of interactive media and not be limited simply to an analysis of interactive videotex. My analysis differs from other studies which have concentrated on technological capabilities rather than on the broader communicational phenomenon implied (i.e., home networking).

A major study prepared by the Institute for the Future (Tydeman et al. 1982) defines teletext and videotex in the following terms:

> Systems for the widespread dissemination of textual and graphic information by wholly electronic means, for display on low-cost terminals (often suitably equipped television receivers), under the selective control of the recipient, and using control procedures easily understood by untrained users. (2)

There are two important features of this definition. First, it is *"medium free"*—implying that videotex and teletext are a "broad class of systems and not a single distinct technology" (2). This broad class includes scrolling or recorded voice, captioning, teletext, narrowband videotex, wideband videotex, personal computers, and personal computers with fully online databases. Second, the definition stresses that this is a mass technology: "widespread dissemination," "low-cost terminals," "easily understood." Elsewhere, the study notes that this is not a "new technology." It represents, instead, a coming together of existing technologies:

> It is a product of the merging of communication systems and computing. It represents a new combination of hardware, such as computers, databases, communication networks, television sets and terminal devices,

and software, including frame design graphics and text, gateways, retrieval mechanisms, and billing. (3)

Modified slightly, this characterization of videotex systems may serve as a basis for describing home networking. The definition is too broad to the extent that it incorporates scrolling or recorded voice, captioning, and teletext. These do not imply an "upstream" interactive link to a host computer or broadcast center. Teletext grants the consumer a measure of control by allowing him to capture specific pages of information as they are cycled through the system, but this control does not imply a link to the broadcast center. In the context of the research presented here, teletext is not a true network technology because it does not allow a central computer to monitor choices which have been made at the user end of the system. On the other hand, the definition is too restrictive because it does not incorporate interactive media. If the definition were expanded to incorporate interactive media and limited to exclude unidirectional teletext information, it would be valid as a characterization of the systems which will make home networking possible.

Rather than define home networking in terms of systems, I prefer to stress its status as a broad social/communicational phenomenon. Accordingly, I define home networking as that social/communicational phenomenon involving the integration of the home into a network of computer-based communications emanating primarily from the corporate sector and effected through the introduction of interactive videotex and interactive media systems into the home.

An operational version of videotex will involve a fairly wide consumer base (roughly 30 percent of homes within a fifteen-to-twenty-year period), terminal decoders provided at low cost (approximately two to three hundred dollars apiece), and procedures for use which do not require any special training. Services to be offered will include information retrieval (news, weather, sports, advertising, directories, how-to guides), transactions (reservations, teleshopping, telebanking), messaging (electronic mail), computing (interactive games, financial analysis), and telemonitoring (home security) (Tydeman et al. 1982, 3). Interactive media systems may operate in conjunction with videotex or independently. When operated independently they will most likely be used in conjunction with cable installations, offering a limited interactive capability which will enable the consumer to select programs, participate in polling (referenda), and respond to video appeals. The systems which will make home networking possible are not based on new technology but on a merging of existing capabilities.

1.3 Markets and Services

From a user's perspective, as well as from that of most research into this topic, interactive systems do not signify the integration of the home into corporate networks. They suggest, instead, a diversified package of electronic services which will be delivered electronically to the home. Research has focused on the assessment of these services rather than on the larger issue of home networking. Although incomplete, the analysis of these services presents a relevant starting point for critical study. Such an analysis must include (1) a study of markets for interactive services (their potentials and limits); (2) a review of the industries providing such services, together with an analysis of emerging ownership patterns, and (3) an analysis of the types of information which will be generated in conjunction with these services (how and where it will circulate, who will use and control it, and to what ends).

The actualization of home networking is contingent upon markets for videotex and interactive media. Because there is no evidence, to date, of consumer or citizen need for these systems, markets will have to be created and cultivated. The process of testing and developing markets for interactive systems is a relatively recent phenomenon. It may be said to have begun in earnest with the trials of QUBE (interactive media) and Prestel (interactive videotex), but a picture has already begun to emerge of likely markets and the services which will cater to them.

When the first videotex systems were developed for the British Postal Service, Prestel was described by British researchers as primarily a publishing technology. The literature it engendered emphasized the impact it would have upon that industry. Smith's 1980 book *Goodbye Gutenberg* suggests this fact in its title, and in the text the potential of videotex technology is compared to "an Alexandria without walls" (316). This conception of videotex was reflected as well in Sigel's 1980 anthology, *Videotext: The Revolution in Home/Office Information Retrieval,* where "videotext"—with an emphasis on *"text"*—is characterized primarily as an "information retrieval" technology.

The first Canadian studies, spawned by Telidon, tended to stress videotex as a publishing technology as well. This was reflected in the title of the first Canadian anthology on the home communications "revolution," *Gutenberg Two.* (Godfrey and Parkhill 1979). But this perception soon changed. Writing in 1979, Madden argued that the definition of videotex would "doubtless be expanded to include functions such as banking, bill paying, teleshopping, electronic mail, opinion polling and telesoftware" (7). The concept of videotex as an "informational" publishing technology changed to one which characterized it as a "transactional" technology. In the short history of videotex development, this transformation was the first important step toward identifying the real market for interactive systems. It marked the moment when the promotion of videotex passed from the hands of engineers and government officials into those of marketing personnel. Successful marketing of videotex will be contingent upon the development of interactive transactional services and not upon the promotion of videotex as an information retrieval or publishing technology. Recognizing this was the first conceptual advance in the selling of home networking.

A number of studies have begun to stress that a marketable videotex system will have to incorporate a package of diverse services. The authors of *Teletext and Videotex in the United States* present this argument in the following terms:

> We also wish to reiterate that even though we examined the merging of the technology and applications in terms of the five generic applications (information retrieval, messaging, transaction, computing, and telemonitoring), we believe that successful videotex systems will have to support several or all of these generic classes of services. The reason for this assumption is that the U.S. society is so diversified in interest that few users will be identical in all respects. Further, the combination of many services within the same system increases its utility since the consumer will find it easier to learn one system than a number of disparate ones. Thus, the consumer is likely to make more use of the system, and, conversely videotex systems are likely to achieve higher penetration of the consumer market. (Tydeman et al. 1982, 169)

A study commissioned by the Canadian Department of Industry, Trade, and Commerce drew similar conclusions about the marketing of interactive teleshopping:

> Few consumers would adopt teleshopping as their sole videotex service. Rather, they will purchase teleshopping as one of a number of services in a videotex package. Adoption of teleshopping, therefore, will coincide with adoption of a complete videotex package. Consequently when we speak of consumers adopting teleshopping, we must consider how other services in the package will influence their perceptions and adoption decisions. For illustrative purposes, a typical videotex package might include these services:
>
> 1. Fire and burglar alarms
> 2. News, weather, and sports information
> 3. Video games
> 4. Travel information (airline, bus, train schedules)
> 5. Making ticket and travel reservations (sports events, concerts . . .)
> 6. Home banking, bill payment facilities
> 7. Information retrieval
> 8. In-house computing [and so on.] (Ash and Quelch 1982, 73–74)

Another obstacle to home networking is the problem of identifying the type of individual or household with both the inclination and the resources to subscribe to a package of interactive services. Initial market research has suggested that videotex services will have limited mass market appeal in the short to medium term (five to fifteen years). The logical market for videotex services, at this stage, will be a category of consumers whom marketers have begun to refer to as the "affluentials." A recent article in a marketing publication described this class and their economic power as follows:

> A growing number of the current affluents are young professional couples. Others are white collar senior management. But many are just two-income households. In total, this 17% of U.S. households earned 50% of the country's total money income last year and currently holds 80% of all securities and 60% of all real estate. By 1990, economist Fabian Linden, executive director of research, the National Industrial Conference Board, estimates that one-third of all personal income will be received by households earning $50,000 or more as compared to 25% now. Yet these households will constitute less than 25% of total households.
>
> What does the trend mean? It means greater concentration of purchasing power. It means a shift in purchase patterns for high ticket products. It means opportunities for new products and services aimed at people with affluence and influence . . . the affluentials. ("Affluentials, a Class Market," *Marketing Communications,* December 1983, 17)

These marketing statistics suggest that by targeting interactive services to a particular "class" market which is small in relation to the entire population, these systems will still be viable because they will reach a disproportionately large concentration of economic power. Moreover, affluential households will tend to be better educated, with an awareness of the power of information in making informed choices, together with some training in manipulating and using it. Because a primary feature of videotex systems is the delivery of enhanced information (whether it be more up-to-date, in-depth, or relevant) to consumers, this will appeal to the affluential consumer. Finally, to the extent that affluential households have been profiled as two-income ones, interactive systems will offer convenience (as, for example, in banking and shopping) to couples who are time-conscious.

Introducing a commercially viable interactive system will be contingent upon promoting it as a transactional technology. A key transactional service will be teleshopping. If interactive systems are to attain their full potential in the future, the videotex industry will have to develop a successful teleshopping capability. Studies suggest that developing this potential will present a major challenge to the videotex industry. To a great extent, consumer resistance to teleshopping is the result of past successes in the field of marketing. So much effort has been devoted to fostering the suburban ritual of the family visit to the shopping mall that it will be difficult for marketers to promote the advantages of shopping at home. Indeed, marketing studies have acknowledged that there exist few, if any, advantages to teleshopping:

> Relative to other non-store methods, teleshopping will likely offer: less assortment, less service and information, greater convenience, higher

> perceived costs, and little peripheral satisfaction difference. Greater convenience is perhaps the only clear advantage, while higher perceived costs and less service are significant disadvantages. (Ash and Quelch 1982, 60)

The lack of inherent appeal in teleshopping correlates with the absence of consumer demand for interactive services in general. This reality presents a challenge to marketers, not an insurmountable obstacle. Their task will be to promote shop-at-home services while taking into account the limits of consumer interest and acceptance of such a capability. Teleshopping will not supplant the consumer ritual of the shopping excursion. Its appeal is located elsewhere, in an existing market for mail-order purchases. Recent trends in the mail-order business may prove to be auspicious for interactive shopping. Tydeman et al. (1982) have noted the following favorable trends in the mail order business:

> Home-based telephone and mail-order purchase of products and services (direct marketing) represent a large and rapidly growing fraction of all sales (Sawyer, 1982):
>
> • More than $110 billion worth of direct sales took place in 1980.
>
> • Direct mail advertising volume increased from $2.87 billion in 1970 to $10.0 billion in 1980.
>
> • Direct response advertising expenditures increased by 14 percent from 1979 to 1980, to nearly $22.5 billion.
>
> • Over 45 percent of the total population (or 85 percent of the adult population) ordered at least one item through the mail in the last 12 months.
>
> • Telephone expenditures for "telemarketing" rose by 15 percent during 1980 to $9.8 billion.
>
> • Mail order is growing at a rate of 15 percent per year, or nearly twice the rate of retail sales. (238)

Ash and Quelch (1982) have analyzed the factors contributing to the growth of mail-order purchases. Among these they have noted the following:

> • increasing costs of gasoline
>
> • increasing participation of women in the work force
>
> • declining quality of store personnel
>
> • increasing number of single and elderly households
>
> • increasing affluence
>
> • increasing quality of non-store business, particulary mail-order, brought about by stepped-up efforts to provide reliable money back guarantees and self-regulation by various associations—improved security and image reduces the risks associated with non-store purchases
>
> • increasing levels of education, consumer awareness, and consumer confidence—more people willing to assume the risks associated with non-store shopping. (47)

Creating a viable interactive shopping capability will be contingent upon successes in capturing and transforming a growing market for mail order purchases; it will not supercede existing shopping rituals, nor does it need to.

A major step toward interactive shopping has been taken with the advent of the much-publicized Home Shopping Network, a cable home shopping capability. After experimenting with a number of formats, the cable industry has finally discovered a viable marketing strategy. The prototype for this service is the Home Shopping Network (HSN) of Clearwater, Florida. HSN-1 began national distribution in July of 1985, and was followed in March 1986 by a more upscale version, HSN-2. Heavily discounted merchandise is offered to the cable viewer, who enters his order by telephone. The Home Shopping Network claims a potential audience of over 8 million homes for HSN-1, with another 2 million for HSN-2. Net sales for the six-month period ending February 28, 1986 totaled $63.8 million dollars (Moorecraft 1986). The success of HSN has prompted the world's largest retailer, Sears, Roebuck and Co., to launch its own network in conjunction with the QVC network. It is estimated that the service could reach 7.6 million cable subscribers (Ivey et al. 1986, 62).

Ash and Quelch, in their study *The New Videotex Technology and its Impact on Retailers in Canada,* have identified two distinct interactive shopping services: a "shop-at-home capability" and a "shopping and product information" capability (1982, 53). Whereas the former has been described as a straightforward "search and order vehicle that consumers could use to purchase goods" (Lane 1984, 17), the latter is primarily a promotional device to be installed on commercial premises in order to "build store traffic" (Ash and Quelch 1982, 53). System operators have begun to refer to this second type of interactive system as "public access videotex," and, to date, over 3,000 public access terminals have been installed in "shopping centres, hotel lobbies, hotel rooms and other public locations in North America" (Lane 1984, 17). At this early stage in the development of interactive services, public access videotex has distinct advantages for system operators because it fosters consumer awareness of videotex without necessitating a terminal for every user. Moreover, research has shown that when public access terminals are properly placed they generate "a formidable amount [*sic*] of page accesses" (Lane 1984, 17). A likely scenario for teleshopping would involve acceptance of shop-at-home interactive media based upon an initial familiarity with in-store public access videotex, coupled with exposure to home shopping through cable. Interactive home shopping would facilitate communication with the broadcasting center by allowing the consumer to request merchandise through a key pad, thus eliminating the telephone. Eventually, television appeals could be supplemented with merchandise listed in a computerized database.

Teleshopping will not appeal to all consumers, nor will it work for all products. Consumers who value convenience will be the natural market for such services. Routinely purchased or well-known products (e.g., brand names) will be most suitable for shop-at-home purchases. Marketing teleshopping will depend upon exploring the special informational potential of interactive systems:

> On the positive side, information providers could make a variety of information packages available to the consumer. For example, IPs [information providers] could provide "helpful hints" on product usage, product comparisons, nutritionally balanced meal menus, and/or weekly specials. Consumers might probably value being able to access such diverse services through a single medium. (Ash and Quelch 1982, 55–56)

In recent years, a picture has begun to emerge of likely fee structures for interactive systems. The problem has been to develop a fee mechanism which will be acceptable to consumers. Initially, the industry favored charges based on a specified rate-per-page of information accessed. Consumer resistance fostered a subsequent strategy whereby rates would be based on total log-on time to the system. Finally, the industry has determined that the most palatable structure for the consumer would be charges based on a "monthly subscription fee plus telephone charges where applicable" (Arlen 1983a, 19). Studies also suggest that "advertising will play a role in subsidizing the cost of consumer electronic services" (Tydeman et al. 1982, 225). Marketers have already begun to develop strategies for advertising on interactive systems (O'Toole 1981). Monthly subscription fees and advertising subsidies will be two features of financing arrangements for interactive systems.

A final trend, somewhat unanticipated until recently, concerns hardware at the user end of interactive systems. A ready market may already exist for interactive services among owners of personal computers (Weissman 1983, 30–31). It is a natural step in the evolution of personal computer use to purchase a modem for communication. Indeed, communicating with other computers and on-line databases may be one of the few useful functions of the home computer for the hundreds of thousands of consumers who have purchased them. In the short term, therefore, users of personal computers will provide a logical market for interactive services. Services developed for this group will eventually expand to include consumers with videotex decoders coupled to adapted televisions.

The following trends will therefore characterize the diffusion of computer networking into the home in the near future. Home networking will be propelled by efforts to market interactive videotex and interactive media services. Although interactive systems will offer some information retrieval capabilities (making this akin to a publishing technology), this will be just one service in a total package which will also include transactions, messaging, computing, and telemonitoring. Transactional services will attract consumers to interactive media and videotex. Chief among these will be teleshopping, which will capitalize on existing trends in nonstore shopping and on the current popularity of shop-at-home cable. The natural market for interactive services will be an upscale class of affluential consumers with purchasing power disproportionately greater than their numbers. Services will be financed in part through advertising, in part by a monthly subscription fee. Users of personal computers will be a natural target group for the first interactive services. These trends have been summarized succinctly in a major marketing publication:

> But the main videotex growth in this country will probably be concurrent with the growth of the personal computer market. While there are currently a quarter to a half million such units in use the figure is expected to reach a million or two by 1985 and to jump even more significantly by the 1990s. As more households come on-line, the potential for videotex services will grow. While they are predominantly informational in nature, industry experts expect that they will become transactional in the future: teleshopping and telebanking will be realities for an admittedly small, though certainly up-scale, segment of households. While the primary users of videotex in Europe are businesses (and the same is true in this country for now), home users should be an important part of the U.S. market by next decade. (Morse 1982, part 1, 24)

1.4 Markets and the Videotex Industry

Identifying general trends in the marketing of home networking is easier than analyzing how industry will organize itself to offer particular services. The videotex industry may be divided into four smaller industries based on the different functions involved in offering interactive services. These include four industry categories: information and service providers, system operators, transmitters, and manufacturers. Each of these broad categories may be further subdivided according to the specific companies active in each area.

In 1981, *Business Week* identified the corporations which were beginning to offer services in conjunction with what it termed "the home information revolution" ("Window On the World: The Home Information Explosion," *Business Week,* 29 June 1981, 74–83). The article noted the participation of major financial (Chemical Bank, Citibank, Merrill Lynch), publishing (McGraw-Hill, New York Times, Readers' Digest, Time Inc., Times Mirror), and retailing (J.C. Penney, Sears, Roebuck, Grand Union) institutions, as well as American Airlines, AT&T, the Associated Press, and the New York Stock Exchange (77). More recently, three joint ventures are experimenting with videotex services: Trinitex, involving Sears and IBM (CBS, an initial partner, withdrew in December 1986); Covidea, a combined effort of Chemical Bank, Bank of America, Time, and AT&T; and CNR Partners, a joint venture of RCA, NYNEX, and Citicorp (Field and Harris 1986, 90).

In Canada, 80 percent of cable subscribers are served by three cable companies: Rogers Cablesystems, Télécâble Vidéotron, and Maclean Hunter. The first of these, whose 1.3 million Canadian subscribers constitute approximately a quarter of the national market, is one of the world's largest cable television companies, with an additional five hundred thousand subscribers in the United States. Télécâble Vidéotron is one of the largest cable television companies in the world, the second largest in Canada, and has approximately six hundred thousand Quebec subscribers. Maclean Hunter is the third largest cable firm in Canada. Operating in Ontario, it has approximately four hundred thousand subscribers ("Canada's Top 10 Multiple System Operators," *Cable Communications Magazine,* 52(11): 35, December 1986).

These companies are developing interactive cable services in both Canada and the United States. Vidéotron recently announced its plans to purchase 100,000 videotex terminals for use with its system (Provost 1984, 9). Both Rogers and Maclean Hunter have been testing interactive cable, primarily through their American franchises (Flaherty 1984, 14). These tests include pay-per-view services, security services, public opinion polling, tele-education, "play" cable, teleshopping, telebanking, and information retrieval (Flaherty 1984, 53–87).

When offering interactive cable services, these corporations will function as both transmitters and system operators. A cable company may also serve as an information and service provider for its own system, participating in all the functions involved in the delivery of interactive services except manufacturing.

Other firms involved in interactive networking in Canada are too numerous to list. Approximately 136 groups and institutions participated as information providers in the Vista trial, a joint venture involving the Department of Communications (Ottawa) and Bell Canada ("Vista—What's Over the Horizon?," *Videotex Canada,* Winter 1983, 27). Telecom Canada, the consortium of Canadian telephone companies, is currently market testing its INET 2000 system. It acts as a value-added carrier, offering standardized menus and simplified procedures for access to gateways and information providers. The network is compatible with both graphic and textual display ("Vista," 1983, 29). Provincial telephone companies such as those of British Columbia, New Brunswick, and Manitoba have also begun to test interactive systems ("On-Line Databases," *Videotex Canada,* Winter 1983, B1–B4; D1–D2; G2–G7).

A key corporate figure in the Canadian videotex industry is Infomart, a joint venture of the Southam Newspaper Group and the *Toronto Star.* When Bell Canada withdrew from managing the Vista trials in 1981, Infomart took over this function. Infomart is currently involved in expanding "the existing Ontario-based Teleguide service to other areas of Canada and the U.S.," increasing the market penetration of the Grassroots service to other provinces of Canada, and, perhaps most important, keeping the Vista system alive by converting it into a "consumer-driven service that will be available primarily in hotel rooms" ("Telidon Content Development Grants," *Videotex Canada,* Spring 1984, 48). Other firms involved from an early date in the VISTA trials as information providers were Tele-Direct and the *Globe and Mail* ("Vista," 1983, 27). In Montreal, *La Presse* has cooperated with Vidéotron to create pages of news and current affairs information. In the summer of 1986 Vidéotron announced an interactive teleshopping service for its subscribers (Durivage 1986).

Videotex market trials have also been initiated in recent years. After years of research and field testing, these ventures tested the commercial viability of interactive services. The industry leaders were Videotex America (a joint venture of Times Mirror and Infomart) and Viewdata Corporation (a subsidiary of Knight Ridder Newspapers) (Arlen 1983a, 15–16). Gateway, the Videotex America trial, began in Southern California in 1984, offering "homebanking, . . . shopping, local news, electronic mail" (Arlen 1983b, 40). The Viewdata trial, Viewtron, offered the same services in Miami, Florida, and was under way in the spring of 1984, with a projected 5,000 subscribers (Arlen 1983b, 40). Both projects aimed to create a national service. They therefore enlisted newspaper publishers in a number of U.S. cities as "co-venturing participants" (Arlen 1983a, 15).

In addition, Videotex America and Viewdata engaged the interest of major financial institutions. Gary Arlen, president of a research and consulting firm which specializes in interactive systems, has described the financial industry's contribution to the development plans of these two companies in the following terms:

> Significantly, each company has collaborated with a group of major financial institutions—lending credence to the belief that videotex systems in the U.S. will develop as joint enterprises of information and financial services (the recent market tests all indicate that banking-at-home is an especially attractive part of the videotex package). Although it's still unclear exactly how these telebanking services will evolve there's no doubt about the commitment of America's leading financial institutions to the concept. (1983a, 16)

Several factors contributed to the failure of these systems (Field and Harris 1986, 83). Primary among them were equipment costs coupled with the costs of the services themselves. It is worth noting that the French have been much more successful in launching their national videotex system, Minitel. Unlike the United States and Canada, France has planned and coordinated the development of videotex at a national level as part of a national telecommunications project. The project was initiated by the DGT (Direction Générale des Télécommunications) in 1977, and was designed to guarantee French participation in the electronics and telecommunications industries of the 1980s. The modernization of the telephone system, which occurred in the 1970s, increased the capacity of the telephone network but led to system underutilization. Videotex systems, which could deliver computer-based information and communication services to the home through telephone lines, were seen as a way of increasing system use. These systems required terminal equipment which, if manufactured in France, would serve as the basis for an indigenous electronics industry. In 1978 the DGT began to coordinate the development of videotex services under the guise of an electronic directory service designed to replace paper directories. The necessary terminal equipment was leased to users below cost, in the hope that increased system use would compensate for this loss.

By the mid-1980s these initiatives began to bear fruit. At the beginning of 1986, over 1.4 million "minitel" videotex terminals were in use in homes and offices throughout France. Approximately 2,000 operational services were available on about 1,200 host computers. More than 34 million videotex calls were made in the months of January and February 1986, with an additional 20 million calls for the electronic telephone directory (Nahon 1986, 2).

The penetration of videotex services in France is without parallel in the western industrial states. The French experience demonstrates that when an interactive videotex system is uniform, presents a single interface to the user, and is offered at a reasonable price, it will have mass appeal. However, it is uncertain when the decentralized and fragmented North American videotex industry will be able to offer a comparable service. The Minitel example suggests that when it does, there will be a market for it.

The alignment of industry for the actual transmission of interactive services into the home remains particularly difficult to foresee. There is no evidence to suggest that in North America channels into the home will be owned publicly and operated by the state. It remains to be determined, however, whether the cable industry or the telephone industry, or a hybrid of the two, will become the primary transmitter of interactive services to the home.

The cable industry may have a technological advantage, given the greater transmission capacity of coaxial cable. However, 98 percent of U.S. cable installations in 1980 were designed for one-way use (Tydeman et al. 1982, 111). The simplest way of refitting cable for two-way use would be to introduce another "return" cable into the home. But this would involve a virtual duplication of the existing system and the cost would be prohibitive. A more complex technological solution would be to send and receive data at different frequencies along the same channel. This option would be somewhat less expensive but would nevertheless require considerable investment (Tydeman et al. 1982, 111). In his study of privacy in cable systems, Flaherty (1983, 13–14) noted the difficulties that the Canadian cable industry was having in raising capital to finance conversion.

The telephone industry is both well and poorly suited to deliver interactive services. Because the phone system is based on switched network technology, it is readily adaptable to information retrieval where the user seeks access to a variety of distant computers. The phone system is technologically deficient, however, due to the channel limitations of paired copper wire. Transmission speeds are too slow for many anticipated videotex services. The modem required to transmit at higher speeds (from 2,400 to 4,800 baud) is in a price range which precludes general use (Tydeman et al. 1982, 114–16). Should the cost of these high-speed modems come down to that of currently used modems in the 300 to 1,200 baud range, telephones would have a decided advantage over cable. One study has predicted that this could occur by 1990 (Tydeman et al. 1982, 117). The telephone industry also has a much larger subscriber base than cable services do. In 1982, cable reached only 28 percent of U.S. homes, whereas the phone network has reached nearly all of them (Tydeman et al. 1982, 6).

In their study of the impact of videotex on Canadian retailers, Ash and Quelch (1982, 38–39) have predicted that the cable industry will dominate the residential market for videotex services. Their conclusions are based on two features of the market for interactive services. First, they note that the cable industry already has an installed base among consumers who have been identified as likely users of interactive services: urbanites who value convenience and who live in areas where conversion to two-way systems will be most economical. Second, building upon the premise that a successful package of interactive services will require photographic image display, they observe that this capability is best suited to coaxial cable. Telephone systems, they suggest (39), will capture business applications.

The study prepared by the Institute for the Future (Tydeman et al. 1982) viewed cable as a "major medium for distributing teletext and videotex services" (113). It was less sanguine in its twenty-year forecast for switched telephone networks, noting that major technological changes would be necessary for them to deliver interactive services to the home. These developments would include a significant reduction in the cost of high-speed modems, coupled with an eventual upgrading of the existing analog telephone network to a fully digital one (117–18). The study argued that optical fiber local loops should appear by the end of the century (118). Replacing copper wire with optical fiber would greatly enhance transmission capabilities.

The process of upgrading the telephone network from an analog to a digital system has already begun. The objective is to transform the telephone system into "one compatible coast-to-coast network that can carry voice and data messages simultaneously over the same line" (Keller et al. 1986, 188). With existing telephone equipment, it is impossible to transmit both voice and data communications simultaneously over the same telephone line. Consequently, people who want the new services have to either rent a second telephone line or use a single line and constantly negotiate who will use the line and for what.

Problems exist at the network level, as well, because the existing telephone network was not designed for data transmission. Using the current system to transmit data would result in slower transmission, higher error rates, and general inefficiency. A number of new designs are currently being tested. The most popular standard is called the Integrated Services Digital Network (ISDN). *Business Week* has noted that eight phone subsidiaries of the seven Bell regional holding companies will soon begin testing the ISDN in the United States. The magazine states: "If the tests pan out, experts say that by 1988 AT&T, the Baby Bells, or both could be offering networks based on the ISDN in metropolitan areas such as New York, Chicago, and Los Angeles" (Keller et al. 1986, 190). Pacific Bell concluded testing such a system in August 1986. Termed Project Victoria, the experiment converted analog telephone signals into a faster digital form. A market testing phase is under way in two California communities (Keller et al. 1986, 191).

A third industry contender has also emerged. A recent study noted that cable manufacturers will soon develop a cable which combines power and optical fiber, delivering both electricity and information to the home:

> Then, most logically, it will be the power companies that will buy the fiber/power cable and lease information-carrying capacity on it to others. At that point, it would seem likely that the economics of running cable into private homes will favor combining CATV, telephone and other services into a single fiber/power cable that would also deliver electric power. ("'Fiber/Power Cable' Power," *Cable Communications,* July 1983, 76)

Such a development appears unlikely in the short term. If it were to occur, its ramifications would be considerable for the cable and telephone industries. It is difficult at this stage to foresee what these would be, or how such an arrangement would be effected.

Based on current evidence and trends, the following scenario reflects industry developments in the area of transmission. Although telephone networks have a consumer base vastly bigger than the subscriber base of the cable industry, this base is not equivalent to the market for interactive systems, which, at this stage, do not have a mass appeal. Marketing research has indicated that interactive services are best targeted to an upscale market of affluent households. These households are to be found in greatest concentrations in urban areas, many of which already have cable. Moreover, research indicates that a successful interactive system must offer a package of diverse services, many of which favor high-resolution display and corresponding high-speed rates of transmission. Not only does cable currently cater to this market, but it is also best suited technologically for high-speed transmission. With proper financing, the cable industry could establish itself as a transmitter—as well as systems operator—for videotex services, at least until optical fiber technology and digital networks develop further. When and if telephone networks decide to upgrade on a large scale, the cable industry will have difficulty matching the financial resources of the telephone companies. The introduction of these systems and of fiber technology in particular would be a long-term process. In the interim, the telephone industry will participate in the business market for videotex systems, a market which currently promises to be more active than the home market. Most studies suggest that the home market will develop out of services inaugurated for business.

1.5 Ownership Patterns in the Videotex Industry

North American society is at the crossroads of a vast social experiment in home networking. As Williams (1974) has noted, early patterns in the introduction of technology have a tendency to become fixed with time:

> . . . the history of broadcasting institutions shows very clearly that the institutions and social policies which get established in a formative, innovative stage—often *ad hoc* and piecemeal in a confused and seemingly marginal area—have extraordinary persistence into later periods, if only because they accumulate techniques, experience, capital or what come to seem prescriptive rights. (147)

Gotlieb (1978, 43–44), in one of the better studies on home networking, identifies three patterns in the implementation of new computer/communications technologies: developments can be "market driven," "technology driven," or "needs driven." Gotlieb argues that, in the past, events in the fields of computers and communications have tended to be both market driven and technology driven. He advocates a needs-driven implementation of technology. "Thus it is reasonable to ask whether the future of computer communications ought to be market driven or technology driven at all, or whether instead developments should be channelled to further socially desirable goals?" (44).

Efforts to promote interactive services have clearly been market and technology driven. The previous analysis suggested a dialectic between technology and market factors in the development of interactive services for the home, and also considered the alignment of industry to foster and exploit both technology and markets. Given Williams' assessment of the longevity of early patterns in the implementation of new technology, it is important to consider how technology, markets, and industry are coalescing at this early stage to introduce home networking.

Promoting home networking on a broad social scale is limited by both available technology and the condition of markets for interactive services. In turn, the organization of corporate institutions eager to participate in home networking has been shaped by opportunities and limits set by technology and markets. A primary market factor in the development of home networking has been the lack of a mass market for interactive services. This development has forestalled the introduction of home networking by reducing any economies of scale which would normally be the result of a mass market for both terminal equipment and services. Consequently, industries concerned with promoting interactive services require tremendous coordination ("Vista," 1983, 27). In addition, considerable capital has been and will continue to be required in order to fine-tune markets and to justify the risks of investment in a limited and uncertain market. Only the largest corporations can raise this capital and justify these risks.

In Canada, the government has played a major role in subsidizing pilot projects, field trials, and market tests ("Telidon Content," 1984, 48). Without state involvement, few corporations would be able to take the risks associated with investing in home networking. Only the largest private-sector corporations, often in conjunction with grants from the public sector, have been able to participate in the development of interactive systems.

Because the videotex industry is divisible into four subindustries (information and service providers, system operators, transmitters, and manufacturers), the evolution of home networking can be understood by analyzing patterns in the ownership and control of these industries. System operators are at the core of efforts to promote interactive services and will most likely be the "driving force" in the growth of these systems (Ash and Quelch 1982, 39). Tydeman et al. (1982) have acknowledged the monopolistic tendencies of system operators in the United States:

> In no case has a U.S. videotex system operator decided to abstain from the role of information provider. To date, it is primarily the largest publishing companies that have participated in providing videotex information retrieval services. This suggests at least the possibility of vertical integration, which might inhibit competition by smaller entities in the electronic information market place. (225)

Ash and Quelch (1982) have noted a similar design in the intentions of the cable industry:

> Cable companies are not yet considered common carriers. They generally see themselves as "multicastors"—packagers of information, entertainment, education, and news, as well as developers of programing involved in generating much of that information. Consequently, the cable com-

panies will likely attempt to control all aspects of cable-videotex from page development to delivery. (24)

Evidently, system operators intend to control all aspects of videotex production. Whether they will be successful in this endeavor is yet to be seen. If they are, they will exercise vertical control over three videotex industries: service and information, transmission, and systems operations. Because a successful interactive system will have to offer a diverse package of services, it is unlikely that a single system operator can provide his clients with all the services required. Most probably, the system operator will retain certain services, such as the telemonitoring (security) capability, and function as a "gateway" to such service and information providers as banks and retailers.

The problem of concentration in the videotex industry is not exhausted in the issue of vertical integration by system operators. In the field of retailing, teleshopping may favor concentration of ownership. Having conducted interviews with a number of industry officials, Ash and Quelch (1982) summarized their findings on the impact of videotex on retailers in the following terms:

> Interviewees generally expect large retailers to reap the greatest benefits from the technology, and hence create a trend toward greater retail concentration. The larger retailers have greater financial and managerial resources to put into videotex experiments and the development of videotex packages. (90)

As noted earlier, the U.S. videotex industry is quickly being colonized by three joint ventures, with Trinitex, Covidea, and CNR Partners representing some of the largest American firms (Field and Harris 1986, 90). The long-term plans of these ventures reflect a new awareness of the market for interactive services. The lack of a mass market for interactive services (particularly in the sense that the market will not cut across class lines) implies that system operators will have to establish national systems catering to a geographically dispersed national market of affluentials. Economies of scale will be achieved if these national ventures are successful.

Currently, there is no evidence to suggest that media monopolies in the broadcasting and publishing industries will be threatened by the emergence of interactive systems. These media institutions are already participating actively in the videotex industry. The videotex industry is emerging as a national network system, as opposed to a dispersed collection of local networks, and so it reinforces patterns of national monopoly control established in broadcasting and mass market publishing.

In his book *Pushbutton Fantasies,* Mosco (1982, 16–66) has reached several conclusions about corporate control of the videotex industry. Mosco reviews the evidence that cooperation and collusion characterize relations among major corporate groups with a stake in the communications industry. He suggests that collusion is routine at the highest level of corporate decision making, the formulation of general corporate policy. An important mechanism of transcorporate consensus is the *interlocking directorate.* Citing a 1980 U.S. Congressional study, Mosco notes that two firms with major interests in videotex, AT&T and CBS, have placed representatives on numerous corporate boards, many of which represent corporations active in the videotex industry (27–29). Citing another report, Mosco demonstrates that the U.S. Congress is concerned about the noncompetitive aspects of interlocking directorates:

> Thus there was substantial opportunity for direct policy discussions and potential understanding among these major competitors afforded by the device of interlocking directorates. Such routes could supply convenient conduits for possible private resolution of public debate between monopoly and competition in the telecommunications industry. (30–31)

Equally significant for the videotex industry is the trend toward monopoly expansion by individual corporations with a stake in interactive systems. Sears, Roebuck and Co. has recently begun to expand its retail and insurance business to incorporate a full range of financial services. This development has been described by *Marketing Communications* in enthusiastic terms:

> In March, 1982, Sears rocked the financial world with the creation of the Sears Financial Network, bringing together the largest retailer, the second largest insurance company, the fifth largest brokerage house and the largest real estate operation in America. It was a grouping of financial might, marketing muscle and distribution potential that would dwarf anything that existed before . . . . The grouping of Dean Witter, Reynolds, Allstate Insurance and Coldwell Banker with the [*sic*] Sears, Roebuck and Company brought together a customer base of 36,000,000 credit and catalog customers (45% of all U.S. households and a distribution network of 831 Sears full service stores, 2,388 catalog outlets, 642 Coldwell Banker offices, 350 Dean Witter offices, 1,950 stand-alone Allstate Insurance sales offices. ("A New Course at Sears," *Marketing Communications,* October 1983, 17)

Although it is not possible to predict which corporations—or mix of corporations—will ascend to dominant positions in the interactive services industry, significant patterns are emerging which permit commentary on the general organization of the industry. First, there will be no videotex industry without the participation of major industry leaders in mass media broadcasting and publishing, telecommunications, electronics manufacturing, finance, and retailing. Second, the introduction of home networking will require a relatively high degree of cooperation among different firms in order to create a marketable package of interactive services. As Mosco's analysis suggests, intercorporate cooperation is not uncommon even at the highest levels of corporate policy making. Third, the promotion of interactive services will run parallel to merger and acquisition trends among corporations currently involved in home networking. Expansion and diversification of the businesses managed by these corporations will enhance the ability of individual firms to deliver a full package of services through the interactive channel.

Home networking, therefore, will emerge as a transcorporate technology. It will be promoted by the largest corporate enterprises in tandem with their agenda for diversification and expansion through mergers and acquisitions. The peculiarities and difficulties of the home networking market will elicit novel forms of cooperation and collusion among corporations. Evidence drawn from power-structure research shows that at the policy level cooperation is routine among corporations which are ostensibly competitors. The first commercially viable networks will be national in scope, catering to a market of affluential households. In countries like Canada, the state will lend active support to the interactive services industry; in the United States, government will cooperate by fostering a climate of deregulation in the communications industry. There will be winners and losers among the "actors" with a stake in home networking, but the hegemony of corporate power in the communication and information industries will not be at risk. Indeed, interactive technologies will present novel opportunities for the expansion of corporate influence and control.

Given the uncertainty of markets, the high risks attending investment, and the large volumes of capital required to establish home networking, it is reasonable to ask whether the value of these systems to industry should be assessed in conventional terms. Indeed, evidence suggests that interactive services embody a qualitatively different type of consumer activity. Interactive technology will deliver a variety of services into the home, but this technology will also deliver back to the service provider, the system operator, and the transmitter very detailed information concerning the state of affairs at the consumer end of the system. In the following sections, this return flow of information is examined in order to demonstrate that plans to exploit this informational resource are central to corporate sector commitment to the development of home networking.

1.6 The Cybernetic Content of Home Networking

In his analysis of the institution of television in the United States and Britain, Williams (1974) has underscored a number of traits which were unique to the evolution of this technology. He notes that unlike previous communications technologies, television and radio were developed without a clear notion of their future content. Television, in particular, was promoted without a clear understanding of its programming content. Williams observes: "*It is not only that the supply of broadcasting facilities preceded the demand; it is that the means of communication preceded their content"* (25). Initially, the electronics industry promoted television as a consumer durable—an appliance not unlike a toaster or an iron. The strategy was to develop another market for electronic hardware.

The development of home networking, an emerging communications technology, is comparable with that of its predecessor in broadcasting. The electronics industry is promoting interactive technology as a hardware product, hoping that interactive systems will create a new market for consumer electronics. Videotex decoders and keypads could be used with adapted televisions; modems and software would be needed for personal computer conversion. The significance of interactive systems for the electronics manufacturing industry should not be underestimated. If videotex terminal equipment were to become widely used (for instance, reaching 25 percent or more of North American homes), it would represent a major market for consumer electronics. The promotion of home networking, therefore, parallels developments in the early sponsorship of television. When adapted for commercial application, both technologies embody important markets for electronic equipment.

Parallels between television and home networking are, however, difficult to sustain. Whereas television was promoted primarily (indeed, almost exclusively) by the electronics industry, home networking is attracting the attention of several industries, including electronics manufacturing, telecommunications, finance, retailing, advertising, mass media broadcasting, publishing, home security, and insurance—not to mention public-sector involvement. Not only do many industries have an interest in home networking, but society itself has much at stake given the magnitude and diversity of corporate interest in interactive systems.

At the core of corporate, transindustry interest in home networking is the content of these systems. Content is manifest in interactive systems at two levels. The first level of content concerns the package of services which this technology will deliver into the home. As noted earlier in the review of markets, industry is still in the process of discovering how to develop a marketable content for interactive systems. At this level, the content of future interactive technology is still uncertain. The situation is similar to that of early television, except that even at this stage it is possible to predict that the service content of interactive systems will be considerably more complex and diversified than that of television. Efforts to create a marketable content for interactive systems will involve careful planning, requiring a systematic and intensive effort on the part of industry.

It would be a mistake, however, to limit an understanding of the content of interactive systems to a formula of content as services. The introduction of interactive services into the home will herald a new era of cybernetic consumerism. This cybernetic content of home networking has been better understood by industry. Every transaction which is executed, and every page of information or service which is delivered, will generate its own electronic (machine-readable) record. The return channel in an interactive system will perform double labor for the interactive services industry: not only will it facilitate consumer requests for services, and thereby stimulate consumption, it will also transmit back to industry much relevant information about consumer demand and consumption. This information will include the consumer's identity, the time and place of consumption (demographics), and product characteristics. This data, combined with other information, will generate an invaluable portrait of consumer activity for marketing purposes. These systems will create a truly cybernetic cycle of production and consumption, because every consumptive activity will generate information pertinent to the modification of future production. Because this information will already be in digital form, cybernetic use of it will be cost-effective. Marketing strategies could certainly become, in large measure, self-adjusting.

Opinion polling will be another major use of the return channel of home networking. The major breakthrough in the use of computers for political polling occurred over twenty years ago with the Simulatics Project (de Sola Pool and Ableson 1961). As the project's title suggests, the computer permitted researchers to do more than process polling data: it enabled them to simulate future voting behavior on the basis of historical data. In an article describing the project, de Sola Pool and Ableson (1961) argued that a convergence of three elements made their experiment possible: "first, a body of sociological and psychological theories about voting and other decisions; second, a vast mine of empirical survey data now for the first time available in an archive; third, the existence of high speed computers with large memories" (183). Concerning the contribution of the computer to simulation techniques, they noted:

> The computer makes possible the handling of this mine of data. More important still, it makes possible the precise carrying out of long and complex chains of reasoning about the interactions among the different processes. In summary, we believe that conditions now exist for use of survey data in research far more ambitious than social scientists are used to. *If it is possible to reproduce, through computer simulation, much of the complexity of a whole society going through processes of change, and to do so rapidly, then the opportunities to put social science to work are vastly increased* [emphasis added]. (183)

In a 1980 study commissioned by the Institute for Research in Public Policy, Pergler has analyzed the return channel of interactive systems with a special focus on "social incasting" (10). This term, as employed in Perlger's text, refers to channels transmitting social or political information to a polling or research center (12). Among the activities classified as social incasting are elections, referenda, opinion surveys, advice, and attitude research (10). There are any number of ways in which political and social polling can be incorporated into home networking. In the case of interactive media systems, opinions may be solicited about subjects treated on television. The prototype for such polling was created in the QUBE trials, where viewers were asked to respond to questions posed in community talk shows. In the case of interactive videotex services, it will require some ingenuity to ensure citizen participation in polling. One possible solution would be to offer special service rates to users who are willing to respond to short surveys as they access a service, or to surveys which are periodically (say once a month) cycled through the system.

For those corporations most active in home networking, providing interactive services will generate returns at two levels. An immediate level of revenues will come from charges assessed to consumers and clients. A subsidiary level of returns will be cybernetic. These second-level returns will be the by-product of the two-way channel of interactive systems and will act as vital feedback to the corporate sector. This information will contribute to the value of interactive systems in two ways. First, it can be collected and sold to other firms for marketing purposes. Second, the information may be used internally to monitor a company's activities, improve its productivity, and evaluate its own markets and strategies. I view these cybernetic returns as the informational commodity of interactive systems. As such, they are different from the service commodity, which, although clearly informational in character, represents a "down stream" flow from information and service providers to the home. Informational commodity here designates this return flow of information to the corporate sector as it becomes increasingly aware of the potential to commodify and privatize this by-product information and thereby secure the financial viability of a home networking industry.

Postindustrialists like Bell (1973) have emphasized the importance of knowledge and information as sources of *surplus value* in modern economies. Although one may disagree with their broad sociological conclusions (for example, that society is evolving into a postindustrial phase with a corresponding transcendence of the social problems associated with industrialization), their insight certainly applies to the fundamental character of interactive systems. Interactive technology will not only produce services for sale to consumers in their homes; it will also produce an informational commodity as a by-product of consumption and deliver it back into the hands of industry. There are no convenient mechanisms for evaluating precisely how much value the informational commodity will add to interactive systems. There can be no doubt, however, that these systems will generate such information as a by-product of electronic transactions and electronic polling and that the informational commodity they produce makes them especially appealing to the corporate sector.

Although the content of home networking is problematic for industry at the level of services, evidence suggests that the latent cybernetic content of these systems is better understood. Indeed, the cybernetic possibilities of interactive systems have been foreseen and eagerly anticipated by the corporate sector for over twenty years.

Writing in the *Atlantic Monthly* in 1964 (shortly after the development of time-sharing computers at MIT and the Simulatics Project), Greenberger noted that electronic data generated in conjunction with the activities of the "computer utility" would be a boon to social research:

> As more and more of these services are perfected, an increasing percentage of the day-to-day functioning of man, the economy, and society will be documented and mechanically recorded in easily accessible form. It will no longer be necessary to conduct costly surveys and door-to-door interviews to acquire data on consumer tastes or investment behavior, at times only to find that the data are inappropriate or anachronistic for the needs of research. Research investigators will specify their precise data requirements and will requisition custom studies from the files of the information utility. The studies will be timely and current and a great boon to analysts and simulators. (67)

This information would have a function more significant than its value to researchers. It would enhance corporate sector "planning" and "decision-making":

> As their use develops, these data studies will be invaluable for corporate decision-making and government planning, to the point where they may be woven into the very fabric of these processes. It is not a mere flight of fancy to anticipate the day when information automatically acquired during the operation of the information utility feeds directly into decision mechanisms that regulate the economy and the activity of companies. (67)

For Greenberger, a computer scientist with a vested interest in time-sharing technology and the perfection of systems research, there are no weaknesses in the scenario he describes. The computer utility is an unequivocal social resource because it will enhance systems planning and control. There is nothing problematic about the ramifications of this control for the distribution (centralization) of power in society because, from Greenberger's instrumentalist perspective, the issue is not power but *rationalization.* The rationalization of society's institutions is axiomatic to his world view. It is an indisputable social goal.

In subsequent years, computer scientists' enthusiasm for the computer utility was thwarted by economic and political realities. The computer utility became politically sensitive in the late 1960s when it became apparent that efforts to centralize the storage of personal files had repercussions for individual privacy. Nevertheless, Greenberger's scenario shows that the cybernetic content of today's interactive systems was fully appreciated over twenty years ago. Elsewhere, this content has been referred to as the informational commodity. The particular significance of this content is its value as input to improved computer-based planning, simulation, and decision making.

Greenberger's commentary reveals an early awareness on the part of industry of the computer utility's potential. Industry saw the computer utility as an institution which would improve the quality of social research and planning by improving input to computer analysis. Computer scientists and social scientists moderated their expressions of public enthusiasm as the project became more controversial. But this did not end efforts to perfect techniques of social management based on the computer. Both private- and public-sector corporations have persisted in their attempts to improve the quality of analysis in computer systems by augmenting and improving the quality of input. In the private sector, efforts have been particularly intensive in the area of marketing research. Researchers are keenly aware that information obtained through polling often contains a bias introduced by the manner in which specific questions are posed (Siwolop 1987). The type of information produced by interactive systems will be more pertinent and reliable because it will concern actual behavior and not intended behavior. Instead of asking an individual how much beer he consumes in a week, it will be possible, in effect, to "count the empty bottles" by determining how much beer he bought.

1.7 Market Research: Current Trends and Practices

The long-term success of interactive systems will hinge upon efforts to develop a viable transactional capability. Current computer-based transactional systems generate electronic records of proven value to marketers. In one of the most comprehensive and authoritative studies of institutional uses of personal information contained in automated record systems, the report of the Privacy Protection Study Commission (PPSC) (1977) acknowledged the value of transactional information to marketers:

> The use of electronically recorded transactions for marketing is a well established practice of a variety of organizations, including retailers and financial institutions. Current market practice is to draw upon the vast pool of credit-card transactions and credit-bureau files, and more generalized sources of demographic information, such as census tracts. Point-of-sale services will dramatically expand the base of electronically recorded transactions that marketers can tap. It will not, however, create a new demand for such information; that demand already exists. (119)

Although these remarks apply specifically to point-of-sale technology located in shopping areas, the insight is relevant to home networking, which may be viewed as a point-of-sale technology removed to the home.

Elsewhere, the commission acknowledged that transactional information, when recorded in electronic form, was frequently recycled and processed for marketing purposes; the results of such analysis could then be sold to other firms. The commission noted that this information was becoming an important "institutional asset":

> . . . the widespread acceptance of credit cards has created vast amounts of recorded information that is extremely useful to marketers. The data bases of credit grantors and credit bureaus, particularly those who have automated their records, have emerged as an important institutional asset. Commercial banks and consumer-finance companies use their records on individuals to screen prospective customers for other commercial enterprises. Credit bureaus refine marketers' mailing lists by weeding out individuals with unsatisfactory credit records. (71–72)

The commission lamented that the public was generally unaware that information generated in the context of one relationship (e.g., transactional) was being applied to decision making in other contexts.

In a recent article entitled "Now Merchants Can Learn More about Charge Customers," *Business Week* described a marketing service currently available to the retail industry:

> Retailers who accept MasterCard and Visa can now purchase profiles of customers in specific neighborhoods through a service being offered by Citibank and Donnelly Marketing Information. For about $2,000, Citibank will prepare 17 reports for those merchants who deposit their MasterCard and Visa transaction tickets with Citibank. The reports make demographic assumptions based on the consumer's address, including the number of people in the household, income, number of children, type of automobile owned, and education level. (19 March 1984, 47)

This example is revealing because it demonstrates the value of a single piece of data when combined with other information for marketing purposes. In this case, a consumer's address is combined with statistical data and analyzed by computer. From this analysis, the marketing service is able to derive facts about the consumer's economic status and lifestyle. The starting point for this assessment is a single piece of information: the client's address.

The Discover Card, a credit card recently introduced by Sears, Roebuck and Co., illustrates quite clearly the value of the informational commodity. Moreover, protests which have accompanied its promotion within the credit industry suggest that the informational commodity will be controversial. Credit-card leader Visa has attempted to discourage merchants from signing on with Sears by claiming that Sears will channel consumer information into their retail branch, thereby undercutting the merchants. Not suprisingly, Sears has denied such an intent (Ellis 1986, 166).

Interactive services will produce such information as a matter of course, but they will also produce a surplus of additional information, recorded in more detail and available on a more regular basis. Given the state of current marketing research, where relatively subtle conclusions may be drawn from limited data, it is difficult to overestimate the value of interactive systems to marketers.

1.8 Transactional Marketing and Behavior Scan

The informational commodity of home networking should be viewed in the context of general trends and crises in the marketing industry. A March 1984 article in *Marketing Communications* theorized about the emergence of a new era in marketing. The article (Harnett 1984) divided the evolution of marketing techniques into three periods. In the initial period or "mass marketing era," the focus was on creating an "image." According to the article, during this period "the media were so powerful, the advertiser essentially wanted to know one thing: *Cost per thousand to reach the masses of people.*" A second era followed in the late 1960s and 1970s, when market segmentation occurred: "Marketers began capitalizing on their brand names by creating a proliferation of line extensions, and variations on the established brand names to attract different people." The key to success during this period was the targeted appeal; emphasis was placed on *"cost per prospect"* (38).

The piece argued that a new era in marketing is emerging based on "cost per sale":

> We now enter a new era in which the goal is zeroing in on the individual buyer and the measurement is cost per sale. Not only can segments be pinpointed more accurately than ever but *response mechanisms can be put in place that quickly determine success or failure* [emphasis added]. This is vital because product life cycles are shorter now than ever before. (38)

It continued by suggesting that this new period be termed the era of "Transactional Marketing," ". . . because it suggests that the proposition to the consumer is less a mood or a status symbol and more *an effort to provoke an action within a limited time frame* [emphasis added]" (38). This new accent on cost per sale suggests that not only will marketing become more cost accountable, it will also become more intensive and manipulative.

Transactional marketing aims not to merely establish a product in the consumer's mind, but to motivate him or her to act upon this stimulus. Neither the practice of nor the theory behind motivational marketing is new to the field. What is new (and this is suggested in the choice of the term "transactional") is the introduction of technologies which will make it possible to improve motivational marketing techniques. The *Marketing Communications* article observed that "the concept of opportunity at the point-of-sale is going through a revision largely as a result of technology" (38). Home networking technology is cited as a force propelling transactional marketing:

> Technology has brought the point of purchase home via home computers, videotext, interactive video systems, catalogues, and 800 numbers. People can now respond quickly to exposure from their homes, abetted by credit cards, with the consequent quick reading results. (38)

Transactional marketing is not a mere theoretical possibility or a vague concept with no basis in current marketing practices and technology. Nor is the announcement of an "era" of transactional marketing simply a product of industry hype. An advanced prototype of future transactional systems has been developed in the United States by a Chicago firm, Information Resources Inc. The system, called Behavior Scan, is available to manufacturers and retailers as a marketing research service ("Shifting Sales into High Gear," *Marketing Communications,* January 1984, 15). Behavior Scan combines data from UPC scanners located in supermarkets, household panels, and targetable media. The service records the purchasing habits of volunteer households and correlates them to household (panel) exposure to television advertising. Consumers are encouraged to participate through the use of coupons which are redeemable at the supermarket. In order for the coupons to be valid, the consumer must identify himself to the retailer as a participant in the marketing trial. UPC scanners then create an electronic record of what the consumer buys. This information is combined with an electronic record of household exposure to advertising and processed automatically by computer in order to assess the effectiveness of advertising.

A *Marketing Communications* article enthusiastically acknowledged the potential of Behavior Scan to create a "real world testing environment" for marketing:

> By combining three unique elements: UPC scanners, household panels and targetable TV, IRI created a real world testing environment for evaluating nearly all of the marketing variables that influence purchasing behavior including advertising, pricing, shelf placement, trade activity and consumer promotions. . . . In addition to collecting total store movement data, IRI has recruited a 2500-household panel in each market, in order to track purchase dynamics at the individual household level. In order to collect purchase information at the individual household level, a panelist must simply show an identification card as he or she checks out. The cashier key-enters the household ID number and all purchases are

> automatically recorded by the scanner and set off in a special computer file. ("Shifting Sales," 1984, 15)

Of particular relevance to the research presented here is the television component of Behavior Scan. Combining Behavior Scan with cable television makes it possible to test the effectiveness of a new ad or marketing campaign. The new ad is targeted to a selected sample of households. Their responses, measured in terms of purchases, are then compared with those of a control group. Behavior Scan allows for a more subtle and flexible experimentation process because it enables marketers to target specific households as opposed to larger demographic units usually defined along geographic lines. *Marketing Communications* notes:

> One particularly unique feature of the Behavior Scan® system is the targetable TV capability which permits the exposure of a selected sample of households to a "test" strategy while exposing a matched set of households in the same market to a "control" strategy. Of special importance is that IRI is able to select a new, distinct sample of test and control households for each commercial test (rather than being locked into a fixed "cable A/cable B" design). ("Shifting Sales," 1984, 16)

The impact of Behavior Scan on marketing will be considerable. When interviewed by *Channels* magazine in 1982, John Keon, a professor of marketing at New York University, acknowledged the revolutionary potential of interactive systems when coupled with marketing techniques based on Behavior Scan. He noted: "The effect of this on market research will be similar to the effect the invention of the telescope had on astronomy. . . . For the first time, we can measure action—the impact of persuasion on consumers" (Koughan 1981/82, 26). In the same article, a spokesman for IRI extolled the unique ability of Behavior Scan to "go through a community household by household and decide what ad we want the consumer to see" (26).

As its name suggests, this marketing service is based on the stimulus/response model of behavioral research. But it is important to emphasize that techniques such as those employed by Behavior Scan represent an advance over traditional uses of behavioral research for marketing. This is the case primarily because interactive technology makes it possible to keep a record of consumer exposure to stimuli in an electronic form which is conducive to automated processing. The importance of these systems to marketing cannot be overstated. As noted in the *Channels* article, Behavior Scan "will provide the manufacturers of consumer products the most accurate information ever available on what motivates a customer to buy" (26).

Given the nature of the technologies involved (i.e., merged systems involving electronic data processing and communications), the future may be appropriately termed the era of cybernetic marketing. The response mechanism to be introduced into the home (or at the point of sale) will transmit to industry an electronic record of the consumer's response to a targeted message. This will allow for an interpretation (processed automatically through the computer) of the success or failure of a promotion or an ad. Because the record of both stimulus and response will be in electronic form, automated processing will be possible.

1.9 Industry and the Informational Commodity

Policy analysis of home networking has been keenly aware of the marketing applications of this technology. In their study *Videotex and Teletext in the United States,* Tydeman et al. (1982, 215–16) have corroborated my findings concerning the value of interactive technology to marketers. Their analysis is significant in three respects. First, they confirm that interactive systems will generate tremendous amounts of information for use in the compilation of consumer profiles. Second, they recognize that different firms will share and exchange data, which is consistent with the conclusions drawn by the Privacy Protection Study Commission about general business practices. Finally, and perhaps most important, they acknowledge that there are few legal or regulatory controls on the transfer of such information.

Concerning the potential of interactive technology to generate information and the modalities of the use of this information, Tydeman et al. argued that

> a profile of consumer spending habits, savings, investments, and credit arrangements can easily be created for all videotex users. Credit ratings and credit card information can be updated daily and available instantaneously to anyone, as the Right to Financial Privacy Act has not been extended explicitly to include electronic funds transfer (EFT). Information service providers and system operators can maintain a complete record of each frame addressed and each transaction performed, thus allowing "customized" direct advertising and solicitation. (216)

Moreover, access to the informational commodity of interactive systems will not be restricted to firms directly involved in the interactive services industry. It will also be available for sale to peripheral industries and groups, as well as to governments, with virtually no legal restrictions on its transfer:

> Under present laws, the communications providers and/or videotex system operators can monitor all information requests by users. This information may be stored, for example, to focus personalized advertising. These patterns, or subsets of these patterns, may be furnished to the gov-

> ernment or sold to credit bureaus, investigation agencies, banks, retailers, political parties, product manufacturers, and so on. (215)

The privacy implications of these additional uses of information will not easily deter marketers in their efforts to fully exploit the informational commodity. Marketers will not forgo the use of this resource on the basis of a privacy value which has been poorly defined and unevenly defended. In anticipation of this emerging resource, spokesmen have begun to defend the "right" of business to utilize the informational commodity.

One justification is that these additional uses of information will benefit business and the consumer alike. Ash and Quelch (1982) have stated:

> The decision of exactly where to draw a line between private and public information is not an easy one. Certainly the consumer has a right to privacy. Yet information about consumer behavior could allow marketers to improve their strategies, increase effectiveness of their promotional costs, and *ultimately provide lower prices to the consumer.* [emphasis added]. (119)

Another strategy has been to deflate opposition by arguing that these uses of information are already standard practice within business and do not represent a substantive change in the status quo: "It is noteworthy that, through such activities as the resale of direct mail lists, information disclosures of consumer purchases already exists as a problem" (119). Tydeman et al. (1982) suggest that it may still be possible to guarantee privacy, but that "privacy protection may evolve as an economic service—the greater the privacy required, the greater the cost to the user" (217). Finally, by way of apparent accommodation to the privacy concern, it will be proposed that information be stored only in aggregate form. Ash and Quelch (1982) argue that

> . . . the fact that this information should be controlled does not imply that it should not be put to productive use. It is conceivable that the government policy could allow data base operators to publish consumer segment information. Thus marketers could receive information about the behavior of groups of consumers. At no time, however, should data base operators be allowed to disclose behavioral patterns of an individual. (120)

This argument assures business access to the informational commodity by circumventing the privacy issue altogether.

The new technologies described above will permit corporations to correlate marketing stimulus to client response at the level, and according to the categories, of their choice. A definite option will be to make these correlations at the level of the individual household. This has provoked concern about the privacy implications of the return channel of interactive systems. But this is not the only level at which information will be analyzed and correlated. Other options include assessing information at the level of the neighborhood, the district, the municipality, the province (or state), and so on. Nor is it essential that a group be delineated according to geographic boundaries. Information generated by interactive systems will make it possible to formulate categories tailored to the interests of industry and government. Groups may be targeted by sex, profession, income, ethnic background, education, purchasing habits, and so on. There is no technological limit to the number or types of categories which may be designated for cybernetic assessment.

This ability to correlate stimulus to response at the group level—however the group is defined—suggests that improved techniques of social control are on the technological horizon. The informational commodity of interactive systems will foster a qualitative enhancement of current practices in marketing and social research. The "laboratory" for social research will no longer be a sample test group whose recorded responses are meant to anticipate those of larger groups. Instead, it will be possible to experiment upon the targeted groups themselves. Interactive technology will assist corporate-sector institutions in their efforts to parcel and subdivide the social fabric into specially constituted market units (religious, political, consumer, special interest, and so on). It will enable the corporate sector to reach these groups with targeted messages tailored to their particular interests and to calculate the effectiveness of these appeals with precision. The introduction of these cybernetic forms of social control, based on a principle of "divide and conquer," will have implications for social policy which clearly cannot be dealt with through the application of existing policy.

1.10 Summary

There is considerable evidence to suggest that home networking will become a reality in the years to come. As interactive technology reaches the current stage of market testing, it becomes less and less essential for research to establish the technological and economic viability of these systems. Rather than question the general feasibility of such a project, the problem for research has been to examine the modalities of an implementation process which is taken for granted. The preceding analysis is both a validation of the thesis that home networking is a technology whose time has come and an exploration of the modalities of the implementation of this technology. The analysis of industry and markets confirms the emergence of home networking as a viable technology in two respects. It demonstrates the extent of corporate-sector commitment to home networking, as expressed by widespread corporate interest and activity and by the actual commitment of capital. The analysis also exposes a new sophistication in corporate-sector understanding of home networking (i.e., specific market strategies, where the market lies, the limits and potentials of this market). These are relatively new developments in the evolution of a technology which has been ongoing for over twenty years. Establishing these developments corroborates the thesis that home networking will be an important component of an emerging infrastructure of computer-based communications. However, only an analysis of particular aspects of home networking (technology, markets, services, industry) can help assess the social significance of this technology.

In the early technological stage of its development, when videotex was the province of engineers and technocrats, a vision of this technology was expounded which vastly overestimated its appeal. In subsequent years, the transfer of this technology from the engineer's laboratory into the home has not proceeded at the pace anticipated by early proponents. Market factors, which were not part of the engineer's initial equation, have intervened and considerably complicated the technology transfer process. By analyzing the interaction of technology and market factors, I have identified a number of significant trends.

Home networking will emerge as the result of efforts to deliver interactive services into the home. Marketing analysis has suggested that the success of these efforts will be contingent upon industry's ability to offer a diverse package of services (multiple services through the same system) with a heavy emphasis on transactional services. Initially, home networking will be targeted to a market of affluential households. Services will be offered through networks which are national in scope. A major factor determining these trends is the absence of a ready mass market for home networking. Each of the trends described above represent a necessary adjustment on the part of industry to the realities of the market for home networking services.

Market factors have an obvious significance for industry. But they are also relevant to my analysis because it is only through an awareness of these trends that one may decipher the specific contours of a future home networking capability. The absence of a ready mass market for interactive systems has specific ramifications for the promotion and implementation of home networking. Arrangements sought by industry to create markets while adapting to market limitations will determine the development of home networking along certain lines as opposed to others. Specifically, difficulties involved in establishing a market for interactive services will have ramifications for the structural organization of the videotex industry. Success will be contingent upon two factors: the investment of tremendous sums of capital, available only to the very largest firms, and a relatively high degree of transcorporate, private- and public-sector cooperation and planning. These aspects of the interactive services industry are important for the formation of public policy because they hint at the structural necessity of a home networking capability which is owned, controlled, and managed by the very largest of today's corporate institutions.

The specter of monopoly concentration and control of the industries providing home networking has been a major concern of policy discourse on new communications technology. For the most part, however, the issue has been discussed in terms of the need to guarantee equal access to these systems to both information and service providers and consumers. Transforming the monopoly threat into a social problem of equal access has itself monopolized the discourse on the social implications of interactive technology. Two prominent examples of this thrust in policy discourse would be de Sola Pool's *Technologies of Freedom* (1983), treating the U.S. problem, and the Canadian anthology *Gutenberg Two* (Godfrey and Parkhill 1979).

Policy discourse has been less aware of and less concerned with confronting what I have termed the informational commodity of interactive systems. The poverty of the current approach, preoccupied as it is with the issue of access, is revealed in the following paradoxical scenario for a future home networking capability: to the extent that policy efforts are successful in resolving the access issue (i.e., by assuring universal access to networks), the specter of computer-based social control also increases. Resolving the access issue will be proportional to the emergence of electronic (automated) techniques of social control, because by fostering a high degree of participation (connectedness) in these networks one also ensures the production of the informational commodity—so coveted by the corporate sector—on a scale conducive to effective social management. In this respect the aspirations of policy makers and the dreams of marketers are similar.

Schiller (1981) provides a more comprehensive understanding of the return flow of information generated by home networking. In *Who Knows: Information in the Age of the Fortune 500,* Schiller, a radical theorist, extensively documents the trend toward privatization of information (47–74). He notes that, although the sale of information is not new to society, certain types of information have traditionally been excluded from the sphere of commodification and privatization and regarded as a social resource. Increasingly, however, these informational resources have become subject to privatization. Schiller focuses particular attention on how public-sector information (whose collection and treatment is subsidized by tax dollars) is subject to de facto privatization as requests for access are increasingly frustrated by the federal bureaucracy. Clearly, this privatization process is at work as personal information, hitherto unaffected by acute forms of corporate-sector interest, is transformed into the informational commodity of interactive systems. The privatization of information has already been documented as a general trend in late capitalist economies. I have therefore focused my analysis not on privatization per se, but on the social use of this information for purposes of control.

One may infer from the preoccupation of policy discourse with the access issue that the informational commodity is simply a marginal by-product of interactive systems, an unanticipated and perhaps troublesome effect of their introduction. I have argued to the contrary. The information commodity is a major structural component of a viable home networking capability, and the ability to use this information may be crucial to the commercial viability of this technology. The value of this information has been understood and appreciated for over two decades—in fact, since the concept of home networking first appeared under the guise of the information utility. I have also demonstrated the structural connection between the development and introduction of this technology and major transformations in the marketing of commodities, oriented, with the assistance of technology, toward establishing a scientific correlation between stimulus and response. It is crucial that an assessment of the social significance of interactive services consider the informational commodity as a major structural component of these systems and make it a focal point for investigation.

Policy discourse has not elaborated on the implications of interactive technology for an augmentation of electronic social control. When the issue has been broached, the problem of social control has been transformed and represented in the context of privacy discourse. This transformation is discussed in the next chapter. A ground-breaking study commissioned by the U.S. Federal Trade Commission in 1981, entitled *Interactive Home Media and Privacy* (Nash and Smith 1981), and Flaherty's 1983 study for the Ontario Ministry of Transportation and Communications, entitled *Protecting Privacy: Data Protection in Two-Way Cable Television Services,* are tentative and early efforts to anticipate the political repercussions of the informational commodity which suggest that some attempt will be made at the policy level to regulate the informational commodity in the interest of privacy protection. The actual form which this regulation will take remains to be determined. In chapter 2 I analyze the legal-regulatory context of information control in North American society as a way of anticipating the types and extent of controls (if any) which will govern information generated by interactive systems. Such controls represent the only major constraint which will be placed on the use and exchange of the informational commodity.

2 The Legal-Regulatory Context of Information Control

2.1 Deceptions in the Privacy Debate

> Organization of markets and exchanges facilitates the determination of predictions and the working-out of calculations which in turn have their effect on immediate production as an attempt to provide continuity and stability, but limitations progressively increased as evident in business cycles and their destruction of time rigidities. The monopoly of equilibrium was ultimately destroyed in the great depression and gave way to the beginnings of the monopoly of centralized state. (Innis 1951, 88)

A struggle is under way today within the western industrial states for the control of computerized information. The expansion of telematic systems of communication is at the technological root of this struggle. The issue of control over computerized information first surfaced as a matter of public policy in the early 1960s. At that time the issue was largely identified with the growing tendency to computerize personal files in both public- and private-sector institutions. A number of public concerns related to government intrusion into the private lives of citizens (e.g., wiretapping and forms of political surveillance) were conjoined with the computer issue and labeled as privacy concerns.

There have been innumerable attempts since the 1960s to establish in law, policy, and social theory the basis for a personal "right" to privacy. These attempts have been tortuous and largely inconclusive.[1] Although it is generally recognized that privacy is a workable concept and a principle worth defending through legal and regulatory mechanisms, there has been little consensus on the actual definition of the idea—whether it is a "right," a "principle," a "value," an "interest," or a "constellation" of competing interests. Of equal significance

1. For a review of problems in defining the privacy interest, see Burns (1976).

has been the problem of grounding the right to privacy in some tradition: anthropological, social, or liberal-political. Without delving into an extensive critique of the privacy literature, several of its misconceptions need to be clarified.

The discourse on privacy argues that the problem is essentially one of establishing an appropriate balance between the demands of citizens for personal autonomy ("the right to be left alone") and the demands of the state for information about its citizens. But, in order to justify the government's appetite for information, privacy theory contends that citizen demand for services (in the context of the welfare state) pushes governments into the collection of personal data. According to this interpretation, the confused and even contradictory demands that the public places upon government are at the core of the privacy issue. If this were an adequate characterization of the problem, an appropriate solution would be to simply educate or rehabilitate the public so that it would be more logical in its relations with government.

There is a contradiction at the core of the privacy issue, but it is not the one suggested in the privacy literature. On the one hand, an ideology of economic and political liberalism—a residuum of "classical" liberal philosophy—posits the autonomous, rational individual as the fundamental unit of the social order and prescribes in law and politics a visible commitment to the individual's autonomy expressed through rights and freedoms. In the classical liberal tradition of "negative" freedom, government cannot interfere arbitrarily in the affairs of its citizens. In conflict with this tradition of political liberalism is the reality of mass, corporate society in a late stage of capitalist economic development. The economic order is no longer characterized by competition among small entrepreneurial capitals, but by forms of concentrated and centralized capital which gravitate toward the monopoly control of markets. As the passage cited at the beginning of this chapter suggests, problems of "continuity" and "stability" find their solution in the monopoly period in the "centralized state." Planning and managing of the social and economic life of the nation is accentuated, expanded, and refined. This has political repercussions in the intrusive character of the state's efforts to maintain stability through management of the populace.

The privacy literature does not acknowledge this conflict between traditional liberal values and the reality of structural change in the social-economic order, a change which compels the development of increasingly bureaucratic techniques of control in order to secure stability. To recognize the problem would underscore the futility of attempts to resolve it in terms of "balance." The conflicting forces are not reconcilable short of a total capitulation at either pole. For political and ideological reasons, outmoded liberal values cannot be debunked, nor can the true dimensions and implications of social change be acknowledged. There are no solutions short of a recognition that certain liberal values are not compatible with current mechanisms to ensure continuity and stability within advanced industrial economies.[2]

It is possible, however, to overstate the case for conflict between liberal values and the centralized state. This formulation of the privacy dilemma leans toward an assumption of genuine social resistance to the expansion of corporate-state powers of control. Although such resistance exists in society today, and occasionally achieves some visibility in the activities of particular groups and forms of so-called deviant behavior, the discourse on privacy was not a staging ground for resistance. To date, the principal result of statutory efforts to guarantee individual privacy rights in Canada and the United States has been to accomplish the contrary.

Rule (1973) has demonstrated how U.S. legislation, ostensibly designed to curtail abusive information practices, has tended to legitimate them. Regarding the Fair Credit Reporting Act (FCRA), he states:

> It is the genius of American liberalism that, when faced with a particularly unconscionable practice by some powerful interest, it regulates that interest in such a way as both to mitigate the sting of the abuse and at the same time to consolidate the position of the perpetrators. The Fair Credit Reporting Act stands in this tradition. (214)

The Privacy Protection Study Commission (PPSC), evaluating the U.S. Privacy Act of 1974, concluded that it was more admirable for its apparent "spirit"—which had been universally ignored by the federal bureaucracy—than for any contribution it had made to the restriction of information practices (1977, 499, 532). Writing in a more critical vein, Weinstein noted in the *Washington University Law Quarterly* that whenever other interests such as "administrative efficiency" and "effective law enforcement" conflicted with privacy interests, "Congress merely chose to sacrifice privacy" (1976, 695).

In Canada, the Privacy Act (Bill C-43) was passed by Parliament in June of 1982 and was proclaimed on July 1, 1983. The newness of the act (hereafter referred to as the Canadian Privacy Act) precludes any analysis of its impact on federal information practices. In most respects it is similar to earlier privacy legislation (i.e., Part IV of the Canadian Human Rights Bill, C-72, passed on July 14, 1977). The legislation is not unlike its American counterpart concerning the conditions under which personal information may be disclosed or exchanged among different agencies of government. Subsection 8 states:

2. It is not my intention here to argue that the liberal tradition of respect for the rights of individual citizens be abandoned. On the contrary, research suggests that an active defense of personal freedoms is more urgent in the emerging data-based society than ever before. I wish to signal, however, (as others have done) that these liberal values are themselves in a period of crisis as they are being adapted to the demands of the economic and political order of late capitalism. For a general discussion of the crisis in liberal political philosophy, see Hallowell (1971) and Jensen (1976).

> (1) Personal information under the control of a government institution shall not, without the consent of the individual to whom it relates, be disclosed by the institution except in accordance with this section.
> (2) Subject to any other Act of Parliament, personal information under the control of a government institution may be disclosed. . . . (8(1)(a))[3]

It then includes a listing by case of thirteen different "exemptions" to the nondisclosure rule (8.(2)). Moreover, this section merely applies to the use of personal information "for the purpose for which it was obtained" or a use "consistent with that purpose" (7(a)).

Such a use of information is simply one which has not been designated at the outset in an agency's description of the personal databases under its administration. This description is published by the Treasury Board each year in the *Personal Information Index.* Referring to a similar provision in the U.S. statute, Weinstein has noted that the effect of the act is not to "significantly restrict interagency transfers; it merely requires agencies to consider in advance how information will be used" (1976, 685). Subsection 18(1) grants to the Governor in Council the authority to exempt some data banks from the act. Schedule II, section 3, lists the federal agencies with an obligation to comply with the act; it does not, however, give any indication of those agencies which are exempt. In the context of the act, therefore, these data banks remain secret. In actual fact, although not required by the act, exempted data banks have been listed in the index. One commentator (Cox 1983) has suggested that the act provides a "degree of protection," while arguing that there is "certainly" room for improvement (43). As is the case with all such legislation since the 1970s, the effectiveness of the Canadian Privacy Act "is very much dependent on the actions of government and the bureaucracy" (43).

In the privacy literature, there is clearly no hidden or sinister agenda whose function is to undermine the very rights of the personal privacy that it claims to protect. But it does not necessarily follow that the protection of personal privacy is the primary objective of policy efforts and privacy legislation. I have found no evidence to warrant the conclusion that public- or private-sector (limited as it is) privacy legislation has significantly altered existing information practices in the United States and Canada. It is possible to attach too much significance to liberal values, inserted as they are in the discourse on privacy. At stake in the privacy debate is not so much the claim to protect the individual from privacy invasions as the establishment of ground rules and limits of acceptable institutional behavior in the context of rapid changes in the technologies of infor-

3. Canadian Laws, Statutes, etc. An Act to Enact the Access to Information Act and the Privacy Act, to Amend the Federal Court Act and the Canadian Evidence Act, and to Amend Certain Other Acts in the Consequence Thereof. 29-30-31 Elizabeth II, Chapter III, 1980-81-82. Ottawa, Canada Gazette. Part III, v. 6, at no. 16, 7 July 1981. (Hereafter referred to as the Canadian Privacy Act.)

mation collection, storage, and transfer. The principal change concerns an augmentation in the cost-effective capacity to collect, store, and transfer information.

2.2 Legal-Regulatory Control: An Outline of Concepts

What follows is an analysis of the legal-regulatory framework for the control of personal files in North American society. The objective is to deduce from this framework the rules and procedures governing the collection, use, and exchange of information generated by home networking. The analysis applies to both public- and private-sector institutions and concerns their documentation practices.

The extent to which specific legislation has been enacted to regulate the use of personal files varies considerably between sectors. Most of the legislation has been directed at public-sector institutions. The rules governing institutional use of personal files have been established in various forums: through adjudication (common law and civil law causes of action, appeals to constitutional law), through legislation (various privacy acts, access to information statutes, and fair credit and reporting legislation), and through self-regulation. Legislation has been effected at different jurisdictional levels, both provincial (state) and federal. The regulation of personal files is directed at four activities: collection, client intervention (access to one's files, procedures for the appeal of file accuracy), internal use, and external disclosure. In the privacy literature, principles of balance and protection are fostered by the creation of guidelines for these processes.

Where legislation is concerned, the modalities of enforcement are significant. These may be external or internal to the institutions involved. External controls may be either subjective or objective. Subjective enforcement implies that the individual client is responsible for policing the institution regarding its use of his files. Objective enforcement designates the situation where legislation creates, in connection with a privacy statute, an external agency with jurisdiction to regulate institutional compliance with the law. Internal enforcement relies upon the institution itself to monitor its compliance with the law. (These categories have been drawn from Hondius 1975, 209.)

2.3 Privacy and Common Law Causes of Action

The consensus among privacy theorists is that common law causes of action are not adequate for privacy protection. Moreover, common law is particularly unsuited to protect privacy in computerized files. From my perspective, common law has not been a factor in regulating personal documentation practices. In the United States, some states have recognized in common law a tort of privacy. The protection of privacy under common law in Canada depends upon litigation brought under torts of trespass, defamation, or breach of confidence. The problem with tort law and the protection of privacy is that a civil wrong against a person or his property must be established as intentional or as an injury arising from negligence. In the case of defamation, for example, the truth of a statement is the basis for litigation. It is relatively easy to see how traditional common law does not apply to the conditions of personal data collection, use, and transfer. A file could be "true" but nevertheless cause injury when used in a different context, or when combined with other data. An individual's privacy may be invaded without causing "harm" as recognized by law. Moreover, it would be extremely difficult to demonstrate the intention to cause harm, or the fact of negligence in the case of routine data practices. Common law, or more specifically, the threat of litigation under traditional common law torts, has not played a significant role in the regulation of computerized files.

2.4 Privacy: The First Attempts at Regulation

In the first important volume in the field of privacy studies, Westin (1967) argued that U.S. courts had gradually been recognizing a principle of privacy rights for some time. In tracing this development, he noted that the historical trend had been toward a concept of privacy based on "personal freedoms," as opposed to an older one based on ownership and property rights (355–56). Although such a trend may have existed, it was not until the 1960s that anyone became interested in examining this tradition in the context of public policy research. In the postwar period, a primary factor provoking public interest in privacy protection was the development of electronic technologies of surveillance and automatic data processing.

Computer scientists associated with early time-sharing systems at MIT had been extolling the potentials of the computer utility since the early 1960s (Greenberger 1968). In the latter half of that decade, a modified version of this concept was proposed within the U.S. federal government. Unfortunately, the initiative was taken by a number of agencies who sought to replace an inefficient, redundant, and decentralized federal data system with a more centralized national data center. These proposals to rationalize federal data systems through centralization and interagency coordination became highly politicized. Despite official claims that the activities of a national data center would not concern privacy but would be devoted primarily to the gathering and processing of statistics, the specter of such a colossal institution, incorporating so much information on American citizens, quashed the proposal in congressional subcommittees (Mowshowitz 1976, 194–95; MacBride 1967).

Against the background of public- and private-sector implementation of electronic data processing in the 1960s, and in the light of political controversies such as the proposal for a national data center, three studies of major importance were initiated by the governments of Canada and the United States to examine the status of data processing activities with respect to privacy in personal files. In Canada, the Department of Communications and the Department of Justice jointly sponsored the Task Force on Privacy and Computers. Their report, *Privacy and Computers,* was completed in 1972. In the United States, the National Academy of Sciences sponsored a detailed study of the trend in both public- and private-sector organizations toward computerizing personal file systems. This culminated in the publication of *Databanks in a Free Society* in 1972 (Westin and Baker 1972). The third study, whose 1973 report was entitled *Records, Computers, and the Rights of Citizens,* was conducted under the auspices of the U.S. Department of Health, Education, and Welfare, and had the most direct bearing upon specific legislation. The input from this report was a major consideration in the writing of the U.S. Privacy Act of 1974. As noted earlier, the Canadian Privacy Act went into effect in 1983, superseding a similar but more limited piece of legislation (Part IV of the 1977 Canadian Human Rights Bill). In contradistinction to the protection of privacy through appeals to common law in the courts, the U.S. and Canadian privacy acts set out to actively regulate public-sector information practices. These acts represent the legal-regulatory framework for the control of personal documentation practices in the public sector at the federal level in both the United States and Canada.

2.5 The Collection Principle

The U.S. Privacy Act of 1974 (5 U.S.C. 552a) imposes some limits and sets some requirements for personal data collection. As noted in the report of the Privacy Protection Study Commission (1977), these measures specify that agencies

> (1) collect only information that is relevant and necessary to accomplish a lawful purpose; (2) collect information to the greatest extent practicable directly from the subject individual; (3) give every individual a Privacy Act Statement at the time individually identifiable information is requested of him; and (4) in certain instances, refrain from collecting an individual's Social Security number and information relating to his exercise of First Amendment rights. (513)

The Canadian statute is similar to its American counterpart:

> 4. No personal information shall be collected by a government institution unless it relates to an operating program or activity of an institution.
> 5.(1) A government institution shall, wherever possible, collect personal information that is intended to be used for an administrative purpose

> directly from the individual to whom it relates except where the individual authorizes otherwise or where personal information may be disclosed to the institution under subsection 8(2).
> 5.(2) A government institution shall inform any individual from whom the institution collects personal information about the individual of the purpose for which the information is being collected. (4; 5(1),(2))

Restrictions, however, are not placed on governmental use of the social insurance number to identify personal data. In 1980, the Conservative government proposed that restrictions be placed on the use of social insurance numbers, but these efforts were defeated before legislation was introduced (Rubin 1983, 10).

Privacy in an Information Society, the published version of the report of the Privacy Protection Study Commission, expressed several reservations about the effectiveness of collection guidelines. The commission noted that the "lawful purposes" clause "seems to have resulted in a *modest* [emphasis added] revision and reduction of data-collection forms, and consequently in a modest reduction in data collection itself" (1977, 513). The report noted that the provision concerning the collection from the individual himself of information pertaining to him "does not appear to have changed practices at all" (513). The Canadian statute is too recent to permit such an evaluation of its performance. In the long run, this will depend upon the federal bureaucracy's interpretation of the law. However, there is nothing in the Canadian statute itself to foster more than a marginal change in data-collection practices, as was the case in the United States.

The U.S. report noted a serious loophole in the act concerning the conditions under which a "privacy act statement" would be issued to an individual. This loophole exists in Canadian law as well. The law stipulates that the statement be issued to the individual when information is collected from him. In the Canadian statute, the individual is to be informed about "the purpose for which information is being collected" when an institution solicits information from him (5(2)). The loophole is that the individual is not informed when information is gathered from a third party (PPSC 1977, 514). Furthermore, the PPSC argued that the statement "tends to state the obvious and does not explicitly spell out other possible uses of information" (514). The same applies to the Canadian statute. In addition, the Canadian Privacy Act overrides the "information from the individual" principle when it recognizes thirteen disclosures from other agencies (8(2)). Information must be collected from the individual, except when it is gathered from other agencies. The Canadian act clearly does not permit collection principles to interfere with status quo practices of interagency exchange and disclosure.

To conclude, collection practices are not significantly affected by North American privacy statutes. These acts do not effectively restrict either the quantity or the types of information gathered by federal agencies, nor are they intended to do so. The force of the laws is directed primarily at restricting abusive, arbitrary collection practices which are not sanctioned by an institution's lawful functions.

2.6 The Access Principle

Assuring a client's right of access to files pertaining to him is a principle recognized in both Canadian and U.S. statutes. In the United States, the rules governing access have been complicated considerably by an overlapping between a previous U.S. statute, the 1966 Freedom of Information Act (FOIA), and the Privacy Act of 1974. In certain cases, it is easier for an individual to obtain access to his files through an appeal to the FOIA than through the Privacy Act (PPSC 1977, 509). In an effort to resolve potential confusion in the Canadian statues, the Privacy Act and the Access to Information Act were legislated together (July 7, 1982). The principal function of access provisions is not to redress or affect the balance of power between individual and institution in any real sense. Instead, these clauses are designed to ensure that decision making based on personal files is founded on "accurate" information (clearly in the interests of both client and institution) and also to reassure the client about the fairness of institutional processing of his files.

According to the Canadian Privacy Act:

> Subject to this Act, every individual who is a Canadian citizen or a permanent resident within the meaning of the *Immigration Act*, 1976, has a right to and shall, on request, be given access to: *a*) any personal information about the individual contained in a personal information bank; and *b*) any other personal information about the individual under the control of a government institution with respect to which the individual is able to provide sufficiently specific information on the location of the information as to render it reasonably retrievable by the government institution. (12(1))

The act establishes procedures for the processing of access requests by clients and creates rules for institutional response (13(1),(2); 14). Guidelines for the correction of files are also set out in the law (12(2)).

Most important, it specifies a number of exemptions to the access provision. These are to be exercised by the head of the respective government institution. They apply primarily to information which may be sensitive in the context of international affairs or defense and to information related to the suppression, prevention, or detection of hostile activities (21), but are equally applicable to information received in "confidence" in the context of intergovernmental affairs (e.g., federal-provincial) (19(1),(2)). Exemptions also include personal information gathered in connection with the activities of an "investigative body" in the context of the exercise of law enforcement activities (22(1)). The act stipulates that the head of a government institution "shall refuse to disclose information that was obtained or prepared by the Royal Canadian Mounted Police while performing policy services for a province or municipality. . . ."(21(2)). Finally, the Privacy Act provides for the exemption of entire data banks designated by the Governor in Council (18(1)). The head of an institution may refuse access to information in an information base exempted under 18(1)(18(2)).

The U.S. statute is similar to the Canadian one in that it contains special exemption privileges for records maintained by the CIA and other law enforcement agencies ((j)(1),(2)). In addition, the law allows for other exemptions applying to law enforcement, investigations, statistics, testing material, and third-party sources of information relating to government employees, contractors, and military personnel ((k)(2),(4),(7);(m)).

In the United States, exemptions apply to "*systems of records* rather than to specific requests for access to specific information" (PPSC 1977, 509). These exemptions must be claimed in advance by an agency for a given information bank under its administration. As a result, some agencies have "made excessively broad claims of exemption" (509). The PPSC noted that this may have been a factor in the low number of access requests made under the act. It also noted that agency rules governing individual access, although in compliance with the law, were often "difficult to comprehend" and that public knowledge of personal information banks was low: "It is doubtful that many people know they exist, let alone how to locate and interpret them" (509). For these and other reasons it was found that the public was not exercising its right of access to the extent anticipated by the law.

In Canada, public awareness of personal information banks has been facilitated through the distribution of the *Personal Information Index* to all post offices, where request forms for access are also available. Under the previous statute (Human Rights Act, Part IV), twenty-two federal information banks were exempt from access (Cox 1977, 27). This is approximately the number exempted by the new act (Rubin 1983, 20). It is too soon to evaluate the record on access requests under the new law. Under the previous statute, 13,081 requests were received between March 1, 1978, and March 31, 1979 (with 2,436 exemptions applied); 8,196 were received from April 1, 1979, to March 31, 1980 (with 1,921 exemptions applied); and 11,350 requests were made between April 1, 1980, and March 31, 1981 (with 3,201 exemptions applied) (Cox 1977, 29).

It is difficult to assess the access provisions of the North American privacy acts because no standard exists for evaluating what access levels should be. In Canada, which has a population of over 20 million, the public is not participating to any significant extent in the law. The average yearly request rate is just under eleven thousand. Moreover, it is likely that most of these requests have come from public-sector workers interested in reviewing their employment files, and do not represent general public participation in the act. This has also been the case in the United States. When files are sensitive, exemptions are applied—extensively, as the figures suggest. Averaged over a three-year period, exemptions were applied in 23 percent of all requests.

The access provisions of the North American privacy acts do not have a major impact on corporate-state information practices. They are somewhat effective in restoring a measure of control in specific cases to individuals who are both knowledgeable and ambitious enough to exercise their rights. As a percentage of the national population, these individuals are small in number. In order for the access provision to dramatically affect information practices, many more people would have to exercise their rights. This would increase the cost of processing such requests and ultimately raise the fee charged for access to personal files, and higher fees would no doubt reduce the number of requests.

Despite the exemptions, the access provisions of the privacy acts do have a positive, but limited, impact on the individual's ability to control information about him in his dealings with public agencies. As a measure affecting the prerogatives of the corporate state regarding the management of mass clienteles, the access provisions are of minor significance. They may oblige agencies to be more cautious, but this caution is directed at accuracy; it does not affect or alter the fundamental mechanism of control, the linking of the individual to his past through documentation. In fact, it should refine the mechanism by improving accuracy and, by extension, improving decision making.

The access provisions in the privacy acts correspond in many respects to the basic objectives of the Canadian Access to Information Act and the U.S. Freedom of Information Act. As mentioned earlier, an overlapping between the Privacy Act and the FOIA in the United States has been a source of confusion. When subjected to analysis, this shared jurisdiction reveals something about informational priorities in the United States. At issue is the value of informational privacy in relation to other institutional priorities. Under the FOIA, it is possible to request records containing personal information pertaining to other individuals. If released, the disclosure of this information could be considered a breach of an individual's informational privacy. Weinstein noted that the FOIA "assumes that the agencies will vigorously assert each applicable exemption because their interests always lie in withholding information" (1976, 708). In reviewing actual practices, Weinstein found that this was not the case: "Unlike national security, the agency has no inherent interest in protecting individual privacy: the incentive for a strong defense is missing" (709). Just as important is Weinstein's comment that when the issue of privacy protection in the FOIA has come before U.S. courts, they have placed a higher value on "utility of release" than on the "privacy interest" (708). He concludes:

> The Privacy Act's various exemptions for the FOIA disclosures and the FOIA's inability to safeguard individual privacy require one of two conclusions: either Congress intentionally subordinated privacy interests to public information interests in every case of conflict, or Congress was wholly ignorant of the complex interaction between the two statutes. There are problems with each conclusion. (716)

Two conclusions may be drawn from Weinstein's analysis. First, privacy protection is clearly not a high priority either within U.S. government agencies or with the courts. Second and more important, this example reveals something about actual priorities. It suggests that there is a measure of consensus among the Congress, the courts, and, to a certain extent, government agencies concerning information policy. As determined from Weinstein's analysis, the basic principle of this policy is that information must be allowed to circulate. The principle is counteracted by institutions with power sufficient to intervene and influence legislation (exemptions) and thereby protect their interests. This occurs on a regular basis in the federal bureaucracy. When no countervailing interest with a power sufficient to the task intervenes, the principle is in effect. This has been the case with the privacy interest.

2.7 The Use and Disclosure Principle

The most decisive category of legal-regulatory controls affecting information practices in the public sector are those related to the use and disclosure of information. In the United States, the PPSC has made a distinction between two primary areas of control governed by two privacy legislation principles: "the use limitation principle" and "the disclosure limitation principle" (1977, 515–21). Use limitation concerns "*internal* uses to which an agency may put information about an individual" (515). Disclosure limitation concerns "limits on the *external* disclosures of information an agency may make" (517). The Canadian and U.S. privacy acts set out to limit use and disclosure in a similar fashion. The Canadian act stipulates that

> [p]ersonal information under the control of a government institution shall not, without the consent of the individual to whom it relates, be used by the institution except
> *a*) for the purpose for which the information was obtained or compiled by the institution or for a use consistent with that purpose; or
> *b*) for a purpose for which the information may be disclosed to the institution under subsection 8(2). (7(a),(b))

The U.S. Privacy Act attempts to limit internal use of personal records by focusing only on those employees with a legitimate right to see a record. It states that, unless an individual agrees otherwise in writing, his records will be available only to "officers and employees . . . who have a need for the record in the performance of their duties" ((b)(1)).

External disclosure is covered in Section 8 of the Canadian act:

> 8.(1) Personal information under the control of a government institution shall not without the consent of the individual to whom it relates, be disclosed by the institution except in accordance with this section.
> (2) Subject to any other Act of Parliament, personal information under the control of a government institution may be disclosed:
> *a*) for the purpose for which the information was obtained or compiled by the institution or for a use consistent with that purpose;
> *b*) for any purpose in accordance with any Act of Parliament or any regulation made thereunder that authorizes its disclosure. (8(1),(2))

Subsection 2 continues by listing eleven additional disclosure exemptions. These include disclosures "for the purpose of complying with a subpoena or warrant . . ."(c); "to the Attorney General of Canada for use in legal proceedings . . ."(d); "to an investigative body specified in the regulations . . ."(e); "under an agreement or arrangement between the Government of Canada or an institution thereof and the government of a province, . . . a foreign state, an international organization, . . . for the purpose of administering or enforcing any law or carrying out a lawful investigation"(f); "to a member of Parliament for the purpose of assisting the individual . . ." (g); "to officers or employees of the institution for audit purposes . . ."(h); "to the Public Archives . . ."(i); "to any person or body for research or statistical purposes . . ."(j); "to any association of aboriginal people . . ."(k); "to any government institution for the purpose of locating an individual in order to collect a debt . . ."(l); and "for any purpose wherein, in the opinion of the head of the institution, (i) the public interest in disclosure clearly outweighs any invasion of privacy that could result from the disclosure, or (ii) disclosure would clearly benefit the individual to whom the information relates" (m).

The external disclosure exemptions in the U.S. Privacy Act are not markedly different from their Canadian counterparts. They include disclosure for law enforcement (b)(7), for statistical purposes (b)(5), for auditing (b)(10), for the National Archives (b)(6), and to either House of Congress (b)(9). The second and third conditions constitute the greatest loophole in the law. The act states that disclosures are prohibited unless "required under section 552 of this title"(b)(2), or "for a routine use as defined in subsection (a)(7) of this section and described under subsection (e)(4) (D) of this section"((b)(2),(3)). Section 552 referred to in (b)(2) is the Freedom of Information Act. The individual does not have to be consulted when the release of a record is required under the FOIA. Release under the FOIA takes precedence over disclosure limitation under the Privacy Act. "Routine use" in the Privacy Act is described as "the use of such a record for a purpose which is compatible with the purpose for which it was collected"((a)(7)). Subsection (e)(4)(D) refers to the listing in the Federal Register for every "system of records" of "each routine use of the records contained in the system, including the categories of users and the purposes of such use"((e)(4)). The PPSC noted that the problem with the routine use provision was in the potential for abuse in the interpretation of the term "compatible." The commission contended that a number of agencies had interpreted it "quite broadly" (1977, 517). On a more critical note, a report prepared for the Ontario Commission on Freedom of Information and Individual Privacy argued that the second and third exemptions in the U.S. act, together with the very general language of the first condition (officers and employees with a "need for the record in the performance of their duties"(b)(1)), "do much to undermine the effectiveness of the protection suggested by the general rule"(Brown, Billingsley, and Shamai 1980, 145).

Legislation governing the internal use and external disclosure of personal files within the public sector represents the principal limitation on the circulation of personal files in North American society. (Legislation pertaining to the private sector is generally less restrictive, less uniform, and focused primarily on the credit relationship; this legislation is reviewed in the next section.) At least two questions are pertinent to an analysis of the use and disclosure provisions of the privacy acts. To what extent have they affected status quo use and disclosure practices within the public-sector bureaucracy? To what extent are status quo practices, past and present, constituted by the relatively free transfer (internal and external) of information contained in personal files? Evidence contained in a number of Canadian and U.S. policy documents suggests that status quo practices of information exchange have not been significantly altered by existing legislation. These documents have also noted that information is routinely transferred on an informal basis both within the public sector and between public- and private-sector organizations.

In Canada, the Task Force on Privacy and Computers noted that more data interchange occurs "than is generally realized by the public" (1972, 30). It also acknowledged difficulties in giving a precise description to the data paths followed by this information. Although many agencies (particularly the large ones) had formal security and disclosure policies, the task force discovered that in actual practice the rules were often "forgotten when communications with persons who can be trusted to handle the information discreetly are involved" (30). It stated: "To a greater or lesser extent, police, credit-reporting agencies, insurance companies, educational authorities and welfare agencies are (amongst other categories) involved in such exchanges, which usually take place without the person involved being informed" (30).

Administrators, the task force observed, are not likely to seek information beyond that required for specific decision making as part of their designated functions. On the other hand, there are few incentives for the administrators to protect the confidentiality of information under their control. This contrasts with the position of statisticians, who, for example, generally must guarantee confidentiality as a condition for receiving information. Most important, the task force observed that administrators were likely to follow the path of least resistance in their dealings with information:

> Furthermore, an administrator will naturally look for the least complicated way of obtaining information. If the information he wants is likely to be embarrassing to the individual concerned, it is often easier to get it from anonymous sources. . . . Mutual administrative convenience therefore leads to the formation of informal exchange networks which permit network members to bypass regulations designed to protect confidential information. (53)

In the United States, a parallel report called *Databanks in a Free Society,* prepared under the auspices of the National Academy of Sciences (Westin and Baker 1972), arrived at similar conclusions:

> We saw that sharing of information about individuals among various organizational realms has been widespread, often goes on unknown and unconsented to by the individual involved, and is sometimes in direct violation of the promises of confidentiality under which he had supplied the information originally. (350)

The study continued:

> Our findings documented the existence of organizational data sharing in violation of law or organizational rules, information "buddy system" exchanges that were either tacitly approved by the record-holding organization, or achieved in clandestine fashion by employees and outside parties. (373)

Both the Canadian and the U.S. reports predate major privacy legislation in their respective countries. Whether irregular data sharing practices still characterize the functioning of public bureaucracies is a moot question, because the privacy acts have succeeded in regularizing most, if not all, of the data transfer activities which had already become common practice prior to the enactment of legislation (PPSC 1977, 532). Moreover, enforcement under both the Canadian and the U.S. privacy acts relies upon the individual citizen. There is no administrative organ responsible for enforcing the acts, as there is in a number of European countries. One may speculate that to some extent the absence of an effective enforcement mechanism encourages informal, nonsanctioned exchanges of information. The acts grant the individual citizen a measure of control over personal information related to him. But society at large cannot have an adequate expectation, based on realistic procedures of enforcement, that the rules designated in the acts can or will be enforced.

The internal use and external disclosure provisions of the Canadian and U.S. privacy acts do not legislate reductions in the transfer of personal information. They may ultimately lead to such reductions, but this is not their focus. If such a development were to occur (and it has not), it would be ancillary to the primary function of the acts, which is to ensure uniform and nonarbitrary exchange of personal files. If an agency has a need to receive or disclose information, it can assure these transfers simply by stating them in advance when it describes the functions of the database. These designated uses must accord with the general purposes for which the particular database was established by legislation. But this need not mean much, because the language of the legislation—"consistent use" (Canadian) and "routine use" (United States)—is sufficiently open-ended to accommodate most purposes. The PPSC noted in its report that the act had little effect on information transfer: "Similarly, as the discussion of the routine-use provision indicated, transfers of information among agencies have only been slightly reduced as a result of the Act's passage" (1977, 532). This may be because the legislation was not designed to reduce transfers, but to regularize existing procedures.

The economic pressures to share information have been noted by Goldstein and Nolan (1975) in *The Harvard Business Review.* These authors argue that certain privacy provisions (but, in particular, restrictions on further use of personal files) may threaten the economic feasibility of personal data bank systems:

> Taking these steps may reveal that some privacy restrictions will render some systems or portions of them entirely unfeasible or unjustifiably expensive, especially those restrictions requiring that all proposed uses of personal data be authorized by the subjects in advance. The need to go back to data subjects to obtain permission for a new use of data already collected may make it almost impossible to expand the usefulness of the system. (70)

In a more critical vein, Rule (1973) has commented upon the political expediency of data sharing among systems of public surveillance:

> No system which develops the capabilities of those studied here can avoid pressures to share those facilities with other interests. The collection of data on a significant share of the general populace, and the maintenance of access to this clientele, are invariably tempting to other persons and organizations, especially those dealing with surveillance and control. The time and expense involved in developing these facilities, and the

> resistance of clients to providing information which may subsequently work to their disadvantage, inevitably place systems which succeed in these tasks at a premium. Given the difficulties posed by large-scale societies for discriminating social control, any system which succeeds in keeping in touch with its people will inevitably be called upon to assist in other functions of control. (318–19)

It merits emphasis that the North American privacy acts do not undermine the economic feasibility of personal data systems in the public sector by placing undue restrictions on additional uses of information. They simply require that further use be anticipated in advance and specified as such (Weinstein 1976, 684–85). As well, disclosure exemptions are both numerous and broad enough to incorporate almost all relevant data transfers. The acts do not seriously jeopardize the routine sharing of data among distinct systems of personal files—the backbone of public surveillance activities as described by Rule. They focus on interdicting nonroutine exchanges of information which, according to their rationale, are inherently abusive or hold the potential for abuse. Abuse is embodied and effected in the rogue activities of individual functionaries in the federal bureaucracy who, for personal gain or through overzealous application of their duties, may disregard normal procedures and thereby undermine an individual's rights. This is clearly the importance of the U.S. Privacy Act, directed as it is at officials with "a need for the record in the performance of their duties" ((b)(1)).

This particular defect in the act was acknowledged by the PPSC. The commission argued that "premeditated abuse," the major preoccupation of Congress in the drafting of the act, was not the problem; instead, abuses "are much more likely to result from the continuing growth of the government's appetite for information about individuals and the use of that information for growing numbers and types of purposes" (1977, 533). The Canadian Privacy Act does not explicitly refer to the problem of "officials" or to its corollary, the control of "premeditated abuse." In this respect, the language of the act is to a certain degree more relevant to the problem it addresses. Its substance, however, is similar to that of the American statute. Neither act aims to reduce the "government's appetite for information."

2.8 Enforcement

The shortcomings of the North American privacy acts are numerous. They fail to reduce to any significant degree the collection of personal data. They do not impose any real control on the use and disclosure of personal files. Their commitment to access is limited to cases where it does not seriously contravene bureaucratic imperatives. No less important is the failure of the acts to legislate effective mechanisms to ensure enforcement of the principles embodied in them. With respect to enforcement, the Canadian and U.S. privacy statutes differ sufficiently enough to warrant separate treatment.

Responsibility for enforcing the rules and general principles incorporated in the U.S. Privacy Act of 1974 lies with the individual citizen. The act represents an improvement over the previous status quo because it recognizes, in principle, that the individual citizen has a legitimate, defensible, legal interest in the information pertaining to him contained in government files. It provides the citizen with both civil and criminal causes of action through which he may protect his interests. In general terms, a basis for civil action exists when an agency fails to comply with some aspect of the act's regulations—and this failure results in an adverse decision for the individual; fails to amend a record upon request of the individual in accordance with the regulations outlined in the law; or refuses to grant access to records in defiance of the law ((g)(1)(A–D)).

In the case of a refusal to grant access, the court is empowered to examine records and, where applicable, to override administrative secrecy. Where an adverse decision has been the result of "intentional or willful" noncompliance with the act, the individual may enter a suit for actual damages against the agency for a sum not less than $1,000. Court costs and legal fees are also assessable to the agency when the individual prevails in his case ((g)(4)).

Violations of disclosure regulations are actionable through criminal actions brought against specific officers or employees of an agency. A cause of action exists when an employee knows of the prohibition against disclosure and "willfully" discloses information. Similarly, a cause of action may be brought against an official who does not maintain a database under his jurisdiction in accordance with the notification requirements for additional use. In both cases, the official found guilty of such a misdemeanor is liable to a fine of not more than $5,000 ((i)(1),(2)).

Under the U.S. Privacy Act, the administration of the act resides with the individual agencies. There is no enforceable administrative supervision at a higher level to ensure compliance. Enforcement lies with the courts, at the discretion of individual citizens who seek to protect their interests. Guidelines for implementing the regulations and agency assistance are provided by the Office of Management and Budget—which itself is not granted any powers of enforcement.

In Canada also, enforcement depends on the individual citizen. A measure of assistance is available, however, in the form of a privacy commissioner. The privacy commissioner is empowered to investigate complaints filed by individual citizens regarding alleged violations of their rights. The scope of these investigations includes violations of disclosure rules and agency failures to grant access, as well as violations concerning the collection, retention, and disposal of personal files as regulated by the act (29(1)).

The privacy commissioner is granted considerable powers while undertaking an investigation. These include the right to summon officials, to administer oaths, to compel the production of documents, and to enter any premise (34(1)). When the privacy commissioner finds that a complaint is "well-founded," he must notify the head of the government institution about the results of the investigation and advise him of any recommendations contained in the report (35(1)(a)). If the agency does not act appropriately, the citizen may appeal to the court for review (35(5)). When the matter before the court concerns a request for access, the burden of proof lies with the agency to demonstrate that the refusal was justified (47). The Canadian statute grants some oversight powers to the commissioner who, by independent initiative, may review the compliance of agencies with the use and disclosure provisions of the act (37). The commissioner may also review the legitimacy of claims to exempt entire data banks from the access provisions of the act (36). The Canadian Privacy Act does not acknowledge a cause of action under the criminal code for violations which may adversely affect citizens' rights. The only penalties recognized by the act are those pertaining to an obstruction of the commissioner in the performance of his/her duties. Fines are not to exceed $1,000 (68(1)(2)).

The success of legislative action is contingent upon mechanisms for enforcement. The Canadian and U.S. privacy acts are similar in their approach to enforcement. In both cases, the compliance of agencies to the provisions of the acts has to be ensured by individuals seeking to protect their own interests. The acts differ, however, in their manner of providing support for the individual's actions. The Canadian statute provides administrative assistance to the citizen by creating a privacy commissioner, but allows for very limited administrative liability and contains no provision for civil or criminal liability. The U.S. act provides for both civil and criminal liability, but provides no administrative assistance.

The question of liability is particularly relevant to both acts because they rely upon citizen initiative to guarantee agency compliance. Incentives are necessary to encourage individuals to register complaints in a process which can be both time-consuming and expensive. These incentives are provided in causes of civil action where an expectation of monetary compensation is sufficient to encourage individual initiative. Similarly, the existence of criminal liability and the potential for criminal causes of action work as an incentive for officials to comply with the law. The pivotal question becomes: Are the incentives provided in the acts sufficient to encourage citizen initiatives?

The provisions for civil and criminal causes of action are considerably stronger in the U.S. act than in the Canadian one. Nevertheless, it has been criticized for not providing realistic incentives. Weinstein (1976) has noted that the original legislation proposed by the U.S. Senate called for the creation of an independent privacy commission with power to "investigate, hold hearings upon and recommend prosecution of agency violations" (692). The recommendation for a privacy commission was not actualized in the final version of the act. More to the point, Weinstein has noted that the incentives for citizen action which supplanted this recommendation are not sufficient to encourage individual initiatives:

> It is unrealistic to expect these remedial provisions to provide adequate incentives for individuals to seek redress under the Act. First, the Act imposes an unreasonably heavy burden on the plaintiff. He must allege and prove that he has suffered an adverse determination in consequence of the agency's violation of the Act, and that the agency acted willfully or intentionally. Most plaintiffs will fail to meet these burdens. The most prevalent threats to privacy stem not from intentional action but from "inadvertent, careless, and unthinking collection, distribution, and storage of records." Requiring proof of willful misbehaviour assures that most abuses will go uncorrected. . . . Second, the Act provides insufficient recovery to stimulate private suits. . . . The $1,000 guarantee is a meager incentive to risk the expense of suing the federal government. . . . Finally, the two-year statute of limitations is unreasonable. (693–94)

The PPSC acknowledged similar shortcomings in the act, arguing that its investigation had uncovered numerous cases where citizen action was an "inadequate tool" for the enforcement of the act. It recommended that for "repeated or systematic" violations, responsibility for enforcement should be delegated to a separate agency with supervisory powers, such as the Federal Trade Commission (1977, 32).

In some respects, the Canadian Privacy Act represents an improvement over its U.S. counterpart. Although the general philosophy of enforcement through individual initiative permeates the act, it also provides for some oversight capabilities on the part of the privacy commissioner. Current plans include staffing a Compliance Directorate with four full-time officials responsible for monitoring agency compliance and submitting reports to the commissioner (Michel Coté, telephone interview, March 1984). The Canadian Privacy Act has clearly benefited in this case from the U.S. experience as outlined in the Report of the Privacy Protection Study Commission. Because the positions have yet to be staffed, it is impossible to evaluate the significance of this development for the conduct of federal information practices.

However, what the act appears to have granted in one area has been denied in another. A report prepared for the Ontario Commission on Freedom of Information and Individual Privacy lamented that Canadian legislation (at that time, the Human Rights Act, Part IV) did not create civil and criminal liability for violations of the act. It argued: "Not one of the rights or duties created by the statute is enforceable by the imposition of a penalty for non-compliance" (Brown, Billingsley, and Shamai 1980, 170). The current act is consistent with the earlier statute in not recognizing criminal and civil liability. In the case of a request for access which has been denied by a government agency, the citizen may receive some satisfaction through the privacy commissioner and ultimately through the courts. (An appeal to the court, however, is justified only if the Commissioner recognizes the legitimacy of the subject's complaint.) The act does not, however, allow the individual to file suit against the government for a violation of his rights as stipulated in the act. A more serious shortcoming is evident in the case of unauthorized disclosures of personal information. Although a data subject may register a complaint concerning unauthorized disclosures of personal information through the privacy commissioner in such instances, he has no legal basis for compensation even if the disclosure affected him adversely.

In this respect, the Canadian act contrasts markedly not only with the U.S. statute but with Swedish and French data acts as well. The Swedish Data Bank Statute of 1973 (c. 289, sec. 1), hereafter referred to as the Swedish Data Act, requires that all personal databases be registered with the government. It stipulates that licenses be granted only to "responsible keepers" who, thereafter, are accountable for managing the data base. If the Inspection Board finds that a keeper has been negligent in complying with the law, fines may be assessed (sec. 24). The Swedish Data Act specifies that violations of regulations are punishable by fines or imprisonment of up to one year (sec. 20). It also acknowledges causes of civil action in order to recover damages from improper collection and use of personal information (sec. 24). The 1978 French Data Act recognizes four offenses punishable by fines of up to 2 million francs and five years' imprisonment (Brown, Billingsley, and Shamai 1980, 161–62).

To conclude, neither the Canadian nor the U.S. privacy acts demonstrate a commitment to enforcement on a par with their European counterparts. The Canadian act could be improved by the incorporation of criminal and civil liabilities which would, in effect, provide it with some "teeth." The U.S. act, in spite of its liability provisions, is inferior to the Canadian statute because it relies entirely on the actions of individuals to enforce compliance. On the other side of the enforcement coin, moreover, are the principles and specific regulations set out in the acts themselves. These do not embody a commitment to the modification or curtailment of bureaucratic prerogatives in the public sector. Clearly, regulations and enforcement are integral to legislation which seeks to redress the power imbalance between individual citizens and the state. The acts restore a measure of power to individual citizens ambitious enough to assert their interests. They do not address the more important issue of expanding corporate power and the role played by personal documentation in this expansion, nor do they propose to do so. It is also relevant that serious questions are already being raised about the credibility of the Canadian Privacy Act. John Grace, the federal privacy commissioner, has noted that the access provisions of the privacy act are being jeopardized by the provinces, who have instructed the federal government, in accordance with one of the exemption provisions of the act, not to release any information contained in federal data banks received from them (Walkom 1984, 1).

2.9 Privacy and the Credit Relationship

The legal-regulatory context of information control in the private sector of North American society is best understood when compared with a number of European countries, primarily Sweden and France. In both countries, governments have regulated through legislation personal data practices in the public and private sectors. In Sweden, the Data Act of 1973 created a Data Inspection Board responsible for licensing and overseeing all automated data banks containing personal files in both public and private sectors (sec. 2). When a license is issued, it is responsible for establishing regulations governing:

(1) collection of data for the file;
(2) automatic data processing procedures;
(3) technical equipment;
(4) operations on personal data in the file that may be done by automatic data processing;
(5) notification of persons concerned;
(6) the personal data which may be made accessible;
(7) dissemination and other uses of personal data;
(8) preservation and weeding out of personal data;
(9) control and security (sec. 6).

As mentioned previously, the act establishes penalties for violations of the regulations and provides for legal redress when citizens are adversely affected by noncompliance. In France, the *Law Concerning Data Processing, Files and Liberties* was enacted in 1978 (French Law No. 78–17, as quoted in Brown, Billingsley, and Shamai 1980, 109). It created an independent administrative authority, the National Commission on Data Processing and Liberties. Similar to the board created by the Swedish statute, the French commission has responsibility for licensing personal data banks, including those in the private sector (Brown, Billingsley, and Shamai 1980, 109).

In North America, there has been no legislative action comparable to the European statutes to regulate private sector uses of personal files. In the United States, the primary attempt to regulate private-sector information practices on the national level has been directed at the credit relationship. The relevant piece of legislation is the Fair Credit Reporting Act (FCRA) of 1970 (15 U.S.C. 1687).

In Canada, there has been no federal legislation to regulate the credit industry. Legislation has been enacted only in some provincial legislatures (Gibson 1980, 113–14). Because the FCRA was the first major piece of privacy legislation to appear in North America and has served as a model for subsequent legislation, it merits review.

The act characterizes its own function in the following terms:

> It is the purpose of this title to require that consumer reporting agencies adopt reasonable procedures for meeting the needs of commerce for consumer credit, personal insurance, and other information in a manner which is fair and equitable to the consumer, with regard to the confidentiality, accuracy, relevancy, and proper utilization of such information in accordance with the requirements of this title.(602(4)(b))

The key terms in this statement of purpose are "reasonable procedures" and "consumer reporting agencies." The latter term designates the particular industry which has been regulated by the act, whereas the former refers to its specific requirements.

A consumer reporting agency is defined as

> . . . any person which, for monetary fees, dues, or on a cooperative nonprofit basis, regularly engages in whole or in part in the practice of assembling or evaluating consumer credit information or other information on consumers for the purpose of furnishing consumer reports to third parties, and which uses any means or facility of interstate commerce for the purpose of preparing or furnishing consumer reports.(603(f))

The act does not regulate the activities of a credit grantor when preparing a credit report for its own purposes. The requirements apply only to agencies which "regularly" prepare reports on individuals for other organizations for purposes of credit, employment or insurance. If a consumer were to request credit from a major retailer, the requirements of the act would not apply if the retailer were to verify credit in-house, using its own staff. Similarly, the act does not regulate the activities of credit grantors: the ways they may use information internally, disclose it to credit reporting agencies, or exchange it directly among themselves.

The requirements of the act place restrictions on the purposes for which a reporting agency may release information. It may be released to the courts. Upon instructions from the consumer, it may be released to organizations which the reporting agency has "reason to believe" intend to use the information for credit, employment, insurance, or license purposes, and to organizations with "a legitimate business need for the information in connection with a business transaction involving the consumer"(604(1)–(3)). The act forbids the reporting of certain derogatory information when it has become "obsolete" (e.g., reporting of bankruptcies fourteen years after the event, and criminal records seven years after conviction) (605). The preparation of "investigative consumer reports" containing information on a consumer's "character, general reputation, personal characteristics, and mode of living" is also regulated by the act ((a)(1)). The consumer has the right to be informed that an investigative report is being prepared about him and is entitled by specific request to a "complete and accurate disclosure of the nature and scope of the investigation requested" (606(a)(b)). However, the individual is not entitled to see the report, but only to be informed of its "nature and scope." Moreover, the act limits the liability of compilers of investigative reports by stating that a violation of section 606 has not occurred if the agency "shows by a preponderance of the evidence that at the time of the violation he maintained reasonable procedures to assure compliance" (606(c)).

Unlike the privacy acts, the FCRA does not stipulate that the consumer is allowed access to his files. It requires that credit reporting agencies "disclose" to the consumer the "nature and scope" of the information in his files (609(a)(1)). In the case of an investigative report, the agency is not required to reveal its sources of information (609(a)(2)). One of the important features of the act is that it entitles the individual to be notified of all recipients of credit reports pertaining to him (609(3)). Unfortunately, it limits this requirement to "the six month period preceding the request," excepting reports for employment purposes, which are limited to a two-year period (609(3)(a),(b)). Where the accuracy of information is contested by the consumer, the act establishes procedures for resolving the dispute. The agency is required to reinvestigate the record and perform any necessary corrections (611(a)). If a dispute is not resolved, the consumer is entitled to present his case and append his statement to the file (611(b)(c)). When a consumer is denied credit, insurance, or employment on the basis of adverse information contained in a credit report, the user of the report is obliged to notify the consumer and give him the name and address of the reporting agency which was the source of the report (615(a)).

Finally, the FCRA creates civil liability for violations of the act. If a reporting agency "*willfully* [emphasis added] fails to comply with any requirement" imposed under the title, the consumer is allowed actual damages, punitive damages "as allowed by the court," and costs of the action (616). Where noncompliance is the result of negligence, the consumer is entitled to actual damages and costs of the action (617). Compliance with the act is to be enforced by the Federal Trade Commission (621). Criminal sanctions govern the "obtaining of information under false pretenses" (619) and "unauthorized disclosures by officers or employees" (620). Both are punishable by fines of up to $5,000 or imprisonment for up to one year, or both (619, 620).

In Canada, legislation regulating the credit reporting industry followed closely after U.S. efforts. In 1971, Manitoba enacted the Personal Investigations

Act, and Saskatchewan, Ontario, Nova Scotia, Newfoundland, British Columbia, and Prince Edward Island all passed legislation between 1972 and 1974 (Gibson 1980, 114). Quebec did not follow suit. In 1975, the Uniform Law Conference of Canada proposed a "Model Act" for Personal Information Reporting. ("The Personal Information Reporting Act," Proceedings of the Fifty-Seventh Annual Meeting of the Uniform Law Conference of Canada, 80–139.) The proposal was designed to create some uniformity among Canadian statutes. To date, no provincial legislature has acted upon the proposal. (Dale Gibson, telephone interview, March 1984). In what follows, the Ontario Consumer Reporting Act of 1973 is reviewed as one example of a Canadian provincial statute governing the credit industry.[4]

Enacted in 1974, the Ontario Consumer Reporting Act incorporates most of the requirements set out in the FCRA. Like the FCRA, the Ontario statute regulates the activities of the credit reporting industry when preparing consumer reports for credit grantors. The requirements for release of consumer reports, the restrictions on storage and release of derogatory information, and the requirements for disclosure to consumers are substantively equivalent to those of the FCRA. The major difference between the two acts consists of the incorporation into the Ontario statute of a "Registrar of Consumer Reporting" (2). The act specifies that reporting agencies must be licensed with the registrar. Accordingly, it specifies conditions for registration and revocation of licenses (4, 5). In essence, these requirements aim to ensure that consumer reports are processed only by legitimate, responsible business interests (4). In contradistinction to the U.S. statute, the creation of a Registrar of Consumer Reporting implies that consumer complaints are first brought before the registrar. The registrar is empowered to order reporting agencies to "amend or delete information, or by order restrict or prohibit the use of any information that in his opinion is inaccurate or incomplete or that does not comply with the provisions of this Act or the regulations" (13(1)).

The Ontario statute also differs from its U.S. counterpart in establishing specific provisions for the investigation of credit reporting agencies by the Minister of Consumer and Commercial Relations (16, 17). Investigations are to be performed by the executive director of the Business Practices Division of the Ministry. If the director determines that a reporting agency is not complying with the act, he may appeal to a judge of the High Court for an order compelling compliance (20(1)). The act contains penalties for both individual and corporate offenders. Offenses are actionable when a person "knowingly furnishes false information," "fails to comply with any order, direction or other requirement

4. Ontario Laws, Statutes, etc. An Act to Control the Storage and Supply of Personal Information for for Rating Purposes. s.o., 1973, c. 97, 2 July 1974. (Hereafter referred to as the Ontario Consumer Reporting Act.)

made under this Act," or "contravenes any provision" of the act (22(1)(a–c)). Corporate officers who "knowingly" concur in such activities are also liable. Fines are limited to $2,000 or one year of imprisonment, or both. When a corporation is convicted of such an offense, fines are not to exceed $25,000 (22(1), (2)).

Other than the registration and investigation provisions of the act, the most significant difference between the U.S. and the Ontario statutes is the partial regulation in the Ontario act of credit grantors. It stipulates that an organization may not "procure from a consumer reporting agency or cause it to prepare a consumer report . . ." unless it "notifies the consumer of the fact in writing before the report is requested . . ." (10(2)). The FCRA requires only that consumers be notified in the case of an investigative report (606).

The deficiencies of North American privacy legislation issue directly from the assumptions about privacy in the credit relationships that underlie them. The acts presuppose that the principal problem, and the one which needs to be addressed in legislation, is the problem of abuse. In this respect, credit reporting legislation anticipates and is consistent with subsequent privacy legislation for the public sector. Abuse exists under the following conditions:

1. An individual or organization collects personal information under the false pretext of the credit relationship and uses it for other purposes.

2. A credit reporting agency willfully or negligently reveals information to persons or organizations that do not have a legitimate need for personal information in the context of a credit, insurance, or employment relationship involving a client.

3. Information is inaccurate or out-of-date.

4. Derogatory information is maintained in personal files for an unfair period of time (fourteen years in the U.S. statute, seven in the Ontario act).

5. An individual is denied access to personal files concerning him.

These abuses are the principal ones regulated by North American privacy legislation.

At least three major criticisms can be made of the credit reporting acts. First, as attempts to regulate private-sector information practices in which the individual is vulnerable, they are insufficient. The European statutes—which require all personal data files in the private sector to be licensed—deal with the problem better. Second, framing the problem in terms of abuse understates and misrepresents issues of informational control in the private sector. Although it is necessary to restore a measure of informational control to the individual in his relationships with organizations, the acts do not question increasing corporate intrusiveness into the lives of citizens, a trend objectionable in itself. Legislative action obscures this issue. Finally, the acts do not address the problem of abuse (their stated objective) adequately, but constitute halfway measures to this end.

The Ontario act stipulates that reporting agencies "shall adopt all procedures *reasonable* [emphasis added] for ensuring accuracy and fairness in the contents of its consumer reports" (9(1)). The FCRA requires that consumer reporting agencies "maintain *reasonable* [emphasis added] procedures designated to avoid violations . . ." (607). The requirements of the acts are not severe and do not impose any strict liability, because there is considerable flexibility in the language of the statutes for interpreting what may constitute a "reasonable" procedure. Moreover, the U.S. act goes a step further to ensure that liability is limited to a small category of violations. It states:

> Except as provided in sections 616 and 617, no consumer may bring any action or proceeding in the nature of defamation, invasion of privacy, or negligence with respect to the reporting of information against any consumer reporting agency, any user of information, or any person who furnishes information to a consumer reporting agency, based on information disclosed pursuant to section 609, 610, or 615, except as to false information furnished with malice or willful intent to injure such consumer. (610(e))

The focus of this provision is the transfer of information. Sections 609 and 610 apply to the disclosure of file information to consumers, whereas section 615 applies to adverse decisions and consumers' right to inquire about sources of information which have affected them adversely. The importance of the provision is to limit the liability of individuals and organizations that have exchanged information with credit reporting agencies, where these transfers have been discovered by consumers as a result of the disclosure provisions of the FCRA. The effect of this provision is to sanction and guarantee transfers of information essential to public surveillance by raising the threshold of liability for those in the business of exchanging personal information.

A major shortcoming of the credit reporting acts is that they do not effectively constrain the collection of personal information by credit bureaus, let alone by credit grantors, employers, or insurers. For example, they do not question the extent to which information gathered in the context of these relationships may be used for purposes other than those concerning consumer eligibility. Information collected in conjunction with credit, insurance, or employment relationships may be applied to operational or marketing decisions essential to the stability and adaptability of an organization; such information may have nothing to do with consumer eligibility, but the consumer is unaware of this. The problem was acknowledged by the PPSC and formulated into a recommendation directed at the insurance industry. The commission recommended that insurance institutions "clearly specify to an individual those items of inquiry desired for marketing research or other purposes not directly related to establishing the individual's eligibility for an insurance benefit . . . (1977,

195). Needless to say, this recommendation was not acted upon in the United States and was not incorporated into Canadian provincial legislation.

The actual intent of credit reporting legislation is best understood in historical terms. Rule (1973) has noted that before the advent of the FCRA, the legality of credit industry practices was uncertain. The industry itself wanted a judicial or legislative statement which would validate their operations. Accordingly, legitimation was the actual agenda of the FCRA:

> At the same time, it [the FCRA] virtually ratifies the position of credit bureaus with respect to the main patterns of their operations. Except for the statute of limitations on derogatory information, it does not constrain the ability of credit bureaus to collect any information which they deem useful. It does not constrain them against selling reports to any of the main customers who now comprise the bulk of their business. True, the Act does require bureaus to maintain "reasonable procedures" to identify buyers of reports and to determine that the sale conforms to the provisions of the Act. But it forbids their selling reports only if they have "reasonable grounds" for believing that the reports will not be used for such purposes. Thus, although the provisions for consultation with consumers will result in certain extra demands on the time of credit bureau staff, and entail extra expense because of the need to verify the accuracy of filed information, *the law should curtail none of the basic credit reporting activities described in the preceding pages* [emphasis added].(215)

Before summarizing the results of this analysis, two additional statutes—one Canadian and one American—merit brief review.

2.10 Some Additional Statutes

The province of Quebec has recently (July 23, 1982) enacted a public-sector information statute. Bill 65 (c. 30), "Loi sur l'accès aux documents des organismes publics et sur la protection des renseignements personnels" (Quebec, Access to Information and Privacy Act), has been introduced in stages since October 1982. Because its major provisions have been in effect only since July 1, 1984, it is impossible to assess the impact of the bill (Lesage 1983). This act merits special mention because it resembles European data acts in one important respect. Chapter IV of the act provides for the creation of a "Commission d'accès à l'information" (103). The jurisdiction of this commission covers both access and privacy requirements of the bill. Its powers are considerably greater than those of the Privacy Commissioner in Ottawa. Specifically, it is empowered to establish the types of information which may be gathered by certain data banks, as well as their security requirements, the categories of persons allowed access to their files, the conditions for their management, and the purposes for which they may be used. Assumptions about abuse and efficient operations found in other North American privacy statutes are shared by the Quebec act. The approach to legislative control differs, however, in that the commission is expected to play an active role in establishing requirements for data banks. In time, Quebec may come to represent a qualitatively different environment for the handling of personal information in the public sector. It should be noted, however, that questions have already been raised about the independence of the commission (Trudel 1983).

In 1978, the U.S. Congress passed the Right to Financial Privacy Act (5 U.S.C. 3400–22, Supp. 1978). It addresses a major category of potential privacy abuse: the demand for personal information maintained by financial institutions. The impetus for the legislation came from the Bank Secrecy Act of 1970. Designed as a measure to assist the U.S. government in its investigations of "criminal, tax, and regulatory matters," the act requires depository institutions to maintain duplicate records of almost all customer transactions" (Davitt 1979/80, 598). Specifically, it "requires that checks and other charges in excess of $100 be microfilmed and retained for five years" (598). Banks discovered that it was less expensive to microfilm all checks than to be discriminating in their compliance with the law. Therefore, as a matter of practice, banks began to keep records of all checks. The Financial Privacy Act became a political necessity in 1976 when the Supreme Court ruled in a seven-to-two decision *(United States vs. Miller)* that Miller had "no legitimate 'expectation of privacy' in the contents of his bank records" (606). Writing in the *Fordham Urban Law Journal,* Davitt has noted: "The effect of the *Miller* decision was to grant the government unrestricted access to a customer-depositor's financial records through administrative subpoenas" (607). The PPSC described April 21, 1976—the day the *Miller* decision was handed down—as "a fateful day for personal privacy" (1977,7).

The Financial Privacy Act of 1978 (FPA) represents a response to the *Miller* decision. It prohibits any federal agency from obtaining personal records unless one of five special procedures is followed. As stated by Davitt (1979/80), these are: customer authorization, administrative summons or subpoena (when records are "relevant to a legitimate law enforcement inquiry"), search warrant, judicial subpoena, and, finally, when no summons or subpoena authority is available to a government official, a formal written request by the official (610–11). In most cases, notice to the customer is required by the act; it establishes procedures for challenging government access (612–13). The act provides for various exceptions to the notification requirements, among which are the following:

> . . . disclosure of information not identified as financial records of a particular customer; examination of records by supervisory agencies in the

> exercise of their supervisory, regulatory or monetary functions with respect to the financial institution; disclosures authorized under the Tax Reform Act of 1976; *information required by federal statute or regulation* [emphasis added]; disclosure of records where the government and customer are parties to litigation . . . ; when the government seeks only the name, address, account number and type of account, or when the government is exercising controls over a foreign account in the United States; when the financial institution itself is under investigation; and finally by grand jury subpoena. (614)

2.11 Summary

This assessment of privacy legislation in North America demonstrates that the philosophies underlying such legislation (the perception of the problem as abuse, together with prescriptions for regulating such abuse through disclosure to consumers, sanctioning of routine status quo practices of exchange, and requirements to establish "reasonable" procedures) are similar in both Canadian and U.S. privacy statutes. On a comparative basis, the Canadian legislation (which has followed U.S. statutes by several years in the cases of both the Privacy Act and the Ontario Credit Reporting Act) has benefited from the U.S. experience. Specifically, it has incorporated better enforcement provisions. The statutes do not rely entirely upon the individual for enforcement but impose a level of administrative oversight. However, the perception of the privacy problem as one of abuse and the fundamental consensus about efficient operations (often involving the collection of more data) as benefiting both parties (client/consumer and agency) are basic to all the legislation that I have reviewed.

The impact of privacy legislation in both Canada and the United States, in both public and private sectors, has not been to curtail the collection, use, or exchange of personal data. Such a conclusion is consistent with the work of Rule (1973) and Weinstein (1976), as well as with the report of the Privacy Protection Study Commission (1977). Moreover, this conclusion is not surprising, because privacy legislation does not seek to curb the prerogatives of bureaucracies involved in public surveillance. Legislation seeks to regularize these practices by eliminating arbitrary practices and sanctioning "routine" collection, use, and disclosure. The individual gets some measure of control primarily through access rights to his personal files. On a societal level, however, legislation does not jeopardize a fundamental category of corporate control administered through forms of public surveillance.

In the next chapter, this understanding of the North American privacy tradition is applied to an assessment of legal-regulatory controls on the informational commodity of interactive systems. The preceding analysis is relevant to such an assessment in two respects. First, on the basis of my review of statutes, it is possible to determine the extent to which existing legislation will apply to the regulation of the informational commodity. Second, the probable thrust of future controls may be determined by examining current efforts to safeguard privacy in home networking in their larger historical context. If controls are to be imposed on the exploitation of the informational commodity, limiting the effectiveness of new forms of public surveillance and social management, these controls will result from the application of privacy policy. In order to analyze home networking in its relationship to emerging techniques of electronic social control, it is essential that these systems be considered in the light of efforts to create privacy policy for interactive systems.

3
Privacy, Public Surveillance, and Home Networking

3.1 The Inapplicability of Current Statutes

For the most part, current statutes do not apply to the informational commodity. They are not relevant to the regulation of personal information generated by interactive systems because they were not enacted for this purpose. This is a function of the North American approach to privacy protection. In Europe, many countries have enacted comprehensive legislation governing the management of databases containing personal files in both public and private sectors, but the United States and Canada have concentrated on the public sector.

Westin (1982) has described the rationale of the U.S. approach (which accords with Canadian efforts) in the following terms:

> Instead of adopting privacy regulations solely for the automated activity (as the Europeans have largely done with their licensing of EDP systems by data protection boards and commissioners), the United States chose to update and expand traditional privacy expectations in these organizational settings, and apply them to all collection and uses of personal information, whether manual or automated. (111)

He continues by noting that U.S. legislative efforts were split into two categories. Public-sector information practices were covered by "broad fair information privacy codes" on the federal and, occasionally, state levels; private sector regulation was less comprehensive and directed at "particular areas of information activity" (e.g., credit reporting) (111).

Westin's commentary is meant as an implicit defense of the North American approach to privacy legislation. The argument contends that European statutes are incomplete because they apply only to automated data banks. The North American approach is assumed to be superior because legislation (when it has been enacted) is directed at regulating the entire database and not just the automated portion of it. This concern with "thoroughness" has not, however, been carried over to legislation for the private sector, where a number of European statutes have been far more effective and extensive.

According to North American statutes, initiatives are undertaken with data banks in the private sector only when there is evidence of abuse, or the potential for it, within a particular industry. To date, only the credit reporting industry meets these dubious criteria. As a result, credit reporting statutes are the only major laws affecting privacy in the private sector. Because interactive systems will be developed by the private sector, it is logical to assess the impact of existing credit reporting statutes on the information generated by interactive systems.

The numerous shortcomings of credit reporting legislation were examined previously. To begin with, the statutes are not directed at controlling the uses of credit information in any comprehensive sense. Instead, their intent is simply to regulate a middleman industry composed of credit reporting agencies. Credit grantors, for example, are not regulated by the U.S. statute. Other shortcomings include weak and ambiguous language concerning "reasonable" procedures, lack of both effective constraints on the types of information collected and real controls on transfer (the agency must simply have reason to believe that its client will use the information for legitimate business purposes), and access provisions which are directed only at the scope of a credit file and not at the file itself. These characteristics of the credit reporting acts have led such critics as James Rule to conclude that the real agenda of legislation was not privacy protection but the protection of an industry.

There are at least two grounds for doubting that these statutes will have a major impact on the informational commodity. First, one must consider the very limited jurisdiction of the statutes themselves, because they regulate only credit reporting "agencies." As was noted earlier, the statutes do not apply in cases where a retailer or another institution processes credit information in-house. Such a scenario is relevant to home networking, because retailers like Sears, Roebuck will be major participants in these systems. The credit reporting statutes will not apply when industries other than credit reporting (e.g., retailers, information and service providers, and systems operators) seek to exploit the informational commodity either through internal use or external sale and transfer. Second, even where the credit reporting industry is involved, the statutes are ineffectual because their hidden agenda is legitimation, not privacy protection. In Canada, the situation is considerably more complex and has a greater potential for abuse, because statutes have not been coordinated at the provincial level and there is no applicable federal legislation.

3.2 Future Trends in Regulation

Because current legislation does not adequately protect privacy in home networking, pressure will build on governments to enact special legislation. It is certainly possible that legislation tailored to the protection of privacy in home networking will be enacted. Although the precise form of such regulation cannot be predicted, it is possible to speculate on the general thrust and orientation of future statutes. Already there are rumblings in policy discourse of the need to guarantee privacy in interactive systems, and privacy discourse is tentatively beginning to explore the options available to policy makers. By combining these early frameworks (options) with an understanding of the tradition of North American privacy efforts, it should be possible to anticipate—with some margin for success—the types of controls which will govern uses of the informational commodity.

It is useful to review the problem of the informational commodity from the perspective of privacy literature. Westin (one of the principal architects of the North American approach to privacy protection) recently described the historical context for the introduction of home networking in the following terms:

> If there was one lesson that was learned during the first part of the computers-and-privacy debate, it was that *personal information is the vital lubricant of the data-based social system we have been building since the 1950s* [emphasis added]. Knowledge of consumer preferences and behavior is central to marketing of goods and services, extension of credit, decisions about employment and insurance, and many other activities. Computerized personal information is equally sought by government for a host of rational and administration purposes such as tax compliance, law enforcement, licensing, program eligibility, etc. (1982, 102)

He continued with the significant revelation that privacy policy had learned to start from the assumption that this information would be used by both business and government:

> When such information exists in automated form, *we have learned to assume that there will be business uses of it and there will be legal obligations created to produce it for government* [emphasis added], unless organizational policies or new legal constraints are developed to keep such data confidential. (103)

Based on this assumption, grounded as it is in twenty years of experience in creating privacy policy, Westin goes on to describe four potential threats to personal privacy posed by what he terms the "two-way household": (1) "improper commercial use by a system operator," (2) "breaches of confidentiality to private third parties," (3) "pressures on subscribers to authorize release of their profile data," and (4) "investigative or litigative access by government" (103).

The modalities of privacy abuse, as so described, indicate the potential uses of the informational commodity. They also suggest where the pressures for these uses will originate: according to Westin's appraisal, in both public and private sectors. By "improper commercial use" Westin refers to additional uses of information similar to those described in the previous chapter—primarily the sale of lists to external organizations for research and marketing purposes. From the perspective of privacy policy, a particularly objectionable practice would be the sale of information containing "derogatory characteristics" (103). What makes these practices objectionable is that the consumer is unaware of them.

The third category of abuse, pressures to release data, is functionally similar to the first, except that the consumer is aware of additional use. In this case, however, the focus of attention is not the system operator but any one of a number of institutions seeking access to personal files. The problem is not the additional use, but the leverage which institutions may employ to make individuals release personal data. The existence of highly detailed data banks will no doubt tempt organizations to use them in decision making. They may be able to circumvent controls by seeking client/consumer approval for release. The individual will be at a disadvantage, because by not complying with an institutional request he risks losing a benefit or having an application denied.

The fourth threat, "investigative or litigative access by government," is self-explanatory. Although this represents an important threat to privacy, it is by no means the only public-sector use of the informational commodity. The potential certainly exists for governments to legislate the release of specific types of information gathered in connection with the routine operation of interactive systems. Such information, which need not be in personally identifiable form, could be useful in such rational activities as social planning and forecasting.

The second category of abuse is less pertinent because it concerns potential for personal information abuse which is not structural in character.

On a macro level, two general options are available to industry, governments, and policy makers. Controls may be instituted by government (i.e., legislated through statutes), or, alternatively, industry may elect to regulate itself. At this time, it is impossible—given the newness of the systems and the relatively recent interest of privacy policy in them—to predict which macro option or combination of the two will prevail. There is momentum supporting both trends.

Flaherty's (1983) study for the Ontario Ministry of Transportation and Communication is a primary example of the self-regulatory approach to privacy protection in home networking. Like other attempts at formulating policy for interactive systems, it is directed at the system operator. Because the cable industry is a likely candidate for system operator in Canada, it is the object of the study. It is relevant to note that system operators (as opposed, for example, to service and information providers) will be the major focus of attention for privacy policy. There are legitimate reasons for this—at least from the perspective of privacy policy—because, as was noted in chapter 1, system operators will resist attempts to reduce their status to that of a common carrier. System operators will be active in as many aspects of interactive systems as they can possibly retain under their control. If this industry model prevails, then the system operator will be the primary repository for the informational commodity. At the moment, cable companies in the United States and Canada are not only active in developing interactive systems but are also highly visible to the public. Consequently, they have become the focus of attention in the privacy debate on home networking.

Flaherty's (1983) report terminates with nine points or conclusions, some in the form of recommendations. Chief among these is the suggestion that the cable companies implement "a code of fair information practices" similar to the one adopted by Warner Amex for its QUBE trial (4). Just as significantly, the report recommends that this code be incorporated into the contractual agreement with subscribers (5). Individuals would then have a legal basis for action (i.e., breach of contract) if the company violated the code. Other recommendations include: (1) that "identifiable data" not be released to third parties without subscriber approval (6), (2) that cable companies "be held ultimately responsible for the integrity and confidentiality of identifiable data flowing through their cable systems to information and service providers" (7), (3) that the Canadian Radio-television and Telecommunications Commission require the privacy code as a condition for licensing non-programming services to cable companies (8), and (4) that legislation be enacted if the cable companies fail to implement a fair information code governing privacy (9) (146–47).

The Warner Amex privacy code mentioned in Flaherty's report (a copy of it can be found in Flaherty 1983, 180–81) contains the following provisions. First, two general commitments are made to the company's subscribers: that the company will explain its "information gathering functions" to them and that it will "maintain adequate safeguards to ensure physical security and confidentiality of any subscriber information" (1; 2). A number of specific guidelines are then outlined concerning the collection of information. Data on individual subscriber viewing or responses is to be gathered only where it is "necessary to permit billing or to render a subscriber services" (3.a). The code states that no other "individualized" information will be collected without advising the subscriber in advance and allowing him the opportunity not to participate (3.b). It also provides that Warner Amex may prepare "bulk (non-individual) data" for use in developing new services and improving existing ones (i.e., for marketing purposes) but may not release this data to third parties without ensuring "that the identity of individuals is not ascertainable from the data provided" (4). The company will refuse requests for information (e.g., from government) unless there is a legal compulsion to comply. Access provisions are included which enable subscribers to "examine and copy any information developed by Warner Amex pertaining to them" (6). Subscriber information is to be retained "for only as long as is reasonably necessary" (e.g., to verify billing) (7). Mailing lists are not to be released to third parties unless consumers are given the opportunity to remove their names from them (8). Warner Amex agrees to comply with all federal, state, and local privacy regulations and to review its code on a regular basis (9; 11). Finally, a very significant provision is included in section 9 which obliges third party companies involved in the trial to comply with the Warner Amex code. Agreements with these firms "shall specifically incorporate this Code of Privacy by reference" (9).

The keystone of Flaherty's report is the recommendation that cable companies voluntarily adopt a "code of fair information practices," for which the Warner Amex privacy code could serve as a useful model. Because the Warner Amex code itself is embedded in the report, it is logical to assess it before analyzing Flaherty's other points, which are designed to complement the code.

Irrespective of its specific provisions, the Warner Amex privacy code represents a significant development in the debate on privacy in home networking. It acknowledges by its very existence that responsibility lies with the system operator to guarantee the security and confidentiality of personal information generated by interactive systems. It also incorporates measures consistent with the North American tradition of privacy protection. The specific provisions of the code set guidelines for the collection, access, use, and transfer of personal data which are apparently commensurate to the problem of privacy protection in interactive systems. For those privacy advocates who favor self-regulation, the code is no doubt adequate, but as a total and definitive set of privacy safeguards, it is deficient and would be considered so by many privacy activists.

As a comprehensive effort to protect privacy in home networking, the Warner Amex code is inadequate in at least two important ways. Its shortcomings are primarily in the areas of enforcement and jurisdiction. These shortcomings are a function of both the self-regulatory status of the code and of its fidelity to the North American approach to privacy protection. Enforcement under the code depends upon the individual citizen. As was the case with controls imposed by North American privacy legislation, the rules set out in the code are enforceable only if individual citizens elect to challenge institutional "abuse" (or their perception of it). In the absence of such actions by consumers, there is no guarantee of compliance with the code. The incentives provided in the U.S. and Canadian privacy acts are not sufficient to encourage citizens to seek action against government for violations of the acts. These weak incentives (causes of civil and criminal action) are still, however, considerably stronger than those provided in a purely voluntary privacy code. Under the Warner

Amex code, the individual has no basis for legal action (civil or criminal) except for breach of contract. Protecting privacy through breach of contract represents a very weak, if not the weakest, mechanism for safeguarding privacy.[1]

The limitations of the scope (jurisdiction) of the code are just as significant. If one considers the complex mix of industries and firms which must participate in order to make interactive systems commercially viable, it becomes apparent that regulating the system operator addresses only part of the problem. In a viable interactive system offering a diverse package of services, information about the consumer would routinely flow into the data files of banks, retailers, information providers, public utilities, and so on. A voluntary privacy code governing system operators would have no impact on the collection, access, use, and transfer of personal files contained in these third-party data banks.

Flaherty's report attempts to improve upon the Warner Amex code by recommending that it be written into the contract between cable company and subscriber, and by supporting section 10 of the code which requires third-party firms to adhere to it. These recommendations are more significant for what they reveal about the inadequacies of the self-regulatory approach than they are as attempts to rectify its deficiencies. There can be no doubt that incorporating a privacy code into a cable service contract will enhance the basis for causes of action under breach of contract. However, even if the code were incorporated, it would be difficult to predict (1) the success of such an action, (2) whether costs and compensation would justify the initiative, and (3) whether such incorporation would have a generalized impact on privacy protection in home networking. A privacy code, even when incorporated into the service contract, represents a very weak safeguard of privacy rights.

Requiring that third-party participants adhere to the code is more

1. The following assessment of court-supported safeguards on privacy was made by the Canadian Government's Task Force on Privacy and Computers in 1973: "Another major reason why a court-oriented approach to privacy may be inadequate lies in the nature of the judicial process. Reliance on the courts alone, either through the development of common law principles or through judicial interpretation of legislative provisions, has a number of drawbacks. Litigation can be costly and time consuming; principles are developed slowly and on a case-by-case basis even where statute law has been enacted; those who are in a position to avail themselves of the courts may not be representative of the social groups most frequently suffering invasions of privacy; courts may lack the necessary expertise to deal with rapidly changing computer technology; the use of the courts is likely to be principally limited to situations involving substantial pecuniary considerations, whereas equally serious invasions of privacy not involving direct monetary loss might never be brought before the courts; and courts are generally oriented to the redress of specific past wrongdoings. Invasions of privacy may in fact be constant, nagging, and often minor, and may frequently occur without the knowledge of the person concerned. There is need therefore for continual surveillance of information systems with a view to preventing these violations, to supplement the court-oriented process of redressing *post-facto* wrongs" (144–45).

impressive for its intended effect than for any actual impact it is likely to have. Flaherty describes this option in the following terms:

> The cable company, in my view, should be held ultimately responsible for ensuring the confidentiality and security of all personal information moving through its systems, even if it decides to extend this liability by contract with other information or service providers on its cable system. (145)

Under such an arrangement, who would be responsible for enforcing compliance? Is it reasonable for the consumer to expect the system operator to monitor and police all third-party participants? I think not. If a third-party company were to ignore fair information practices, would the system operator bring it to court for breach of contract? It seems unlikely that the system operator would so vigorously defend the consumer's privacy interest as to initiate legal action on his behalf. Or would the system operator simply break its agreement with the third-party firm? Such an action could involve the system operator itself in a breach-of-contract suit. Another alternative would be for the system operator to simply not renew agreements with third parties. Is it reasonable to expect the system operator to adopt such a measure when it may be against its own economic interest?

Recommending that third-party participants adhere to a code of fair information practices acknowledges that the problem extends beyond the system operator. Making the system operator ultimately responsible for the activities of third parties is unrealistic. It represents a patchwork attempt to keep the self-regulatory approach alive in the face of evidence that it does not solve the problem.

The other major option for privacy policy would be legislated controls. In a 1982 article in *Datamation,* Westin reviewed the state of privacy legislation in the United States. He noted that although pressure was building for a federal statute, most legislative activity had been concentrated at the state level. Illinois and Wisconsin were among the states with privacy statutes for interactive systems. Legislation was still pending in California and New York. In addition to these initiatives, Minnesota's cable commission had also established privacy rules governing municipalities and their licensing of cable franchises. All the statutes, whether enacted or pending enactment, were directed at the cable industry. Westin's description of the Wisconsin statute is a relevant indicator of the statutes' scope and tone. He notes:

> The second main provision [the first concerns an on-off device which would control the flow of signals back to the system operator] forbids any person, without the consent of the subscriber, to monitor the subscriber's cable equipment or the use of it, except for service purposes; to provide anyone with information that discloses any "aspect of behavior" of the

> subscriber, of members of the household, including "individual habits, preferences, or finances"; or to "conduct research that requires the response of the subscriber or a member of the subscriber's household, except by mail or personal interview," unless the subscriber is notified in writing before the research begins and "at least once a month" while the research is being conducted. (106)

The statute establishes penalties for violations of the act which enhance enforcement. These include fines of up to $50,000 for a first offense and up to $100,000 for subsequent offenses. Moreover, "damages and injunctive relief can also be awarded to injured parties who sue under the law" (106).

Of special significance are the controls, found in all the statutes, governing both the collection and the disclosure of information to third parties about subscriber viewing habits and preferences. The Illinois Cable Television Privacy Act forbids the cable operator to "give anyone, without express written consent, a list containing a subscriber's name; to 'conduct research for any purpose'; or to 'install or maintain a home-protection scanning device in a dwelling' " (Westin 1982, 104). The rules established by the Minnesota Cable Commission are equally firm with respect to disclosure. The ruling of the commission states: "No information on the viewing habits of a subscriber obtained by monitoring a subscriber's transmission can be sold or otherwise made available to any party other than the operator and its employees unless authorized by the subscriber" (Westin 1982, 111). Westin notes that legislation pending in New York state parallels the Warner Amex code in most respects. The code is strengthened, however, by creating a legal basis for the consumer's expectation of confidentiality (1982, 106).

Subsequent to the Westin article, California's statute went into effect (January 1, 1983). Flaherty (1983) has noted that the act forbids "a cable company from providing 'any person with any individually identifiable information regarding any of its subscribers, including, but not limited to, the subscriber's television viewing habits, shopping choices, interests, opinions, energy uses, medical information, banking data or information, or any other personal or private information without the subscriber's express written consent' " (102).

A pattern appears to be emerging whereby the provisions of the Warner Amex code serve as a model for legislation at the state level. This is particularly true with respect to limitations on both additional use and the disclosure of personal information to third-party firms.

It requires no special insight to appreciate the impact of these statutes on uses of the informational commodity. At first glance, they appear to challenge the very existence of the informational commodity as a basic component of home networking systems. When considered in relation to the legal-regulatory tradition of North American privacy legislation, these controls seem to be substantial, if not severe. Noticeably absent, for example, is the seemingly endless list of exemption provisions to disclosure rules found in the North American privacy acts. At least two questions are relevant to an assessment of the impact of these statutes on the informational commodity. First, do the statutes represent a permanent trend in the regulation of interactive systems? Second, what is the precise impact of these controls on uses of the informational commodity?

The existence of these statutes does not automatically imply that they will serve as a model for future legislation. Current legislation has emerged at a particular stage in the evolution of home networking characterized by the introduction of interactive media systems, precursors to more technologically sophisticated interactive videotex. The statutes repeatedly refer to "viewing habits" and subscriber "responses" (e.g., to television polling)—characteristic of attempts to regulate interactive media technology. The California statute is a notable exception to this trend, with its catalog of different types of information (shopping choices, energy uses, and so on) which the cable company may not disclose to third parties. The later introduction of this statute obliged lawmakers to consider emerging videotex systems along with interactive media capabilities. In all cases, however, these statutes precede the introduction of a widespread, fully operational version of home networking with a full complement of interactive videotex services.

The situation was markedly different for earlier privacy legislation. In the case of both credit reporting legislation and the public-sector privacy acts, controls were introduced to regulate generalized practices which had already become institutionally entrenched. Privacy legislation for home networking is unique because it has been created in advance of the introduction of technology in order to anticipate the data activities of system operators. Because these systems are not yet operational, the privacy interest has not come up against an entrenched commercial interest with a substantial stake in opposing legislation. This situation explains the sporadic success of efforts to create privacy controls for interactive cable. It does not automatically follow, however, that the controls contained in these statutes will become the standard for future privacy.

Taking into consideration the unique anticipatory status of current legislation, it is possible to draw two very different conclusions about future privacy policy for home networking. On the one hand, judging from existing legislation, the trend in the future may be toward equivalent, if not more demanding, legislation. On the other hand, when the same information is considered in a larger historical context, there may be reason for skepticism. Based on past experience in establishing privacy policy, it is reasonable to expect that opposition to safeguards will grow as interactive systems reach a commercially viable stage. It is then that the privacy interest will meet substantial corporate-sector opposition. Accordingly, the current phase is not typical of a future climate for privacy policy. The real test for the privacy interest will commence when interactive systems become fully operational. The evidence presented in chapter 1 suggests that resistance will be formidable, given the mix of participating industries, the current trend toward transactional marketing, and the inherent value of the informational commodity itself.

The North American tradition of privacy legislation does not suggest that substantial constraints will be placed on institutional use of the informational commodity. It is useful to consider briefly how industry is opposing the privacy interest at the current stage. For the most part, industry resistance has materialized in the argument that legislation should follow abuse and not attempt to anticipate it. When challenged by privacy activists, industry has countered that a wait-and-see attitude is preferable to an unconsidered reflex to regulate before a problem emerges. Westin (1982, 106) has noted that industry spokesmen prefer a federal privacy initiative to a continuation of independent legislation at the state level. There are two reasons for this. First, it would be much easier for industry to comply with one set of uniform regulations. Second, of all the government levels involved, the federal government is the least likely at this date to expedite the passage of privacy legislation. Political pressures have been registered sporadically at the municipal and state levels, but they have not consolidated at the federal level. Given the current vogue for deregulation in the U.S. government, particularly regarding the communications industry, cable operators are no doubt at ease waiting for a national privacy initiative. By postponing the inevitable, cable operators improve their strategic position in the privacy debate. Once the required hardware for an interactive capability is in place, it will be difficult to impose controls requiring an upgrading of technology or a restructuring of systems. Moreover, waiting until a clear picture of the videotex industry emerges will enhance the effectiveness of lobbying efforts. No longer will the cable industry stand alone in its opposition to privacy constraints, because other participating firms will also have a stake in the informational commodity.

An important barometer of trends in the regulation of home networking is the position of privacy policy itself. Although preliminary and tentative, the contours of a privacy policy for interactive systems have begun to emerge. Westin's (1982) commentary is replete with the cautious language and apparent reverence for "balance" which have reduced much privacy policy to an exercise in legitimation. Historically, a key to the pruning of the privacy interest has been the introduction of fine-grained distinctions in the privacy debate. In Westin's scheme, "commercial" and "research" uses of personal information are opposed to governmental requests for access. He urges that the problem of privacy abuse be considered separately according to these distinct cases, and notes that in most cases in which the government has requested access to data files the courts have upheld this right. Westin suggests that few controls are necessary at this point to set rules for government access, presumably because the record has demonstrated that when the government sets its sights on information it is usually able to compel its release. The types of minimal controls he recommends are similar to those set forth in the Right to Financial Privacy Act, described in the previous chapter.

A further distinction applies to his consideration of private-sector, commercial, and research uses of information. Westin contends that the debate on these secondary uses may be divided into two camps, the "free market" school and the "prohibitory" approach. A discursive sleight of hand enables him to associate the so-called free market approach with a defense of personal freedoms and the prohibitory approach with its antithesis, which he suggests is the tyranny of the regulators. He states:

> The free market view argues that the decision on secondary uses of subscriber data should be left to the individual viewer, not dictated by the privacy advocates. There would be no penalty for declining, but *a monetary benefit might be provided to those who agree to have their names supplied* [emphasis added] for contact in areas of their consumer or social interest. In my view, this kind of option should be open to subscribers, since the core of privacy is the individual's right to choose what he or she reveals to others. (112)

The Wisconsin statute described above is a particular target of Westin's scorn and disapproval. In his opinion, it resembles the Swedish approach, which he contends is characterized by a "paternalism" that denies individuals their freedom of choice:

> This represents the kind of paternalism and overkill that American privacy protection has not previously engaged in. It resembles the Swedish data protection board's ban on computer dating services, which the Swedes reject as too self-revealing for people to be allowed to choose. In the United States, we let people select such services, and the law steps in only if blackmail or other abuses are found to be taking place. (112)

Clearly, these statements by a prominent spokesman for the policy establishment are more in line with the North American privacy tradition than with early statutes at the state level in the United States. Moreover, they suggest how policy discourse itself will seek to compromise and undermine a strong defense of the privacy interest. Tydeman et al. (1982) proposed that privacy be offered as an "economic service" to those who could afford to pay for it. Westin hints at a similar possibility, but modifies the formula very slightly to create positive economic incentives for those prepared to sell their privacy for an economic benefit. In the first case, individuals are penalized if they cannot muster the resources required to defend their privacy, whereas in the second they are rewarded for forfeiting it.

Both options are objectionable. They both degrade and reduce privacy to the status of a mere commodity which, as such, may be subject to quantification and exchanged for the valueless trinkets of a consumer society. In the future, the market for privacy will be controlled and dominated by the corporate sector. Of the two proposals, Westin's represents the greater threat to a long-term preservation of privacy rights. Under a benevolent facade of individual choice, Westin's proposal effectively guarantees the availability of the informational commodity for secondary uses. Moreover, Westin's plea for freedom is grounded in the distorted proposition that an individual somehow exercises his freedom when he sells off the particulars of his private life for bureaucratic consumption.

Attaching an economic value to privacy, whether it be cast in terms of rewards or punishments, will be a crucial component of strategies designed to circumvent privacy interests and assure commercial exploitation of the informational commodity. In addition to the two examples cited above, Flaherty (1983) has acknowledged a similar "reward" formula which would provide for the release of personal information to third-party firms:

> One plausible hypothesis is that individuals who are determined to have a full range of interactive services without regard to their personal privacy can choose freely to sell identifiable personal information about themselves to cable companies for whatever purposes, in return for a fee or a reduction on standard charges for such services. Cable companies would of course be reluctant to pay consumers for information about themselves, if the data are only being used in aggregated form. (34)

This problem has also received attention in contexts other than that of home networking. A 1977 article in *Science* entitled "Computers in Banking and Marketing" proposed the following scenario as a way of circumventing the privacy interest:

> Realizing that complete security is not possible, we could perform an interesting social experiment. Suppose two electronic funds transfer systems were offered with identical financial systems capabilities. System 1 is very secure and devotes much of its efforts to protecting the privacy of customers; this expense is passed through in-service charges to users. System 2 has easy access for people or organizations who may use the data in the files for any purposes they wish (since the blackmail potential for this is nil, presumably the users would be doing marketing research). Customers are paid a premium for using system 2 and the commercial interests pay the costs of operation. This scheme has the interesting prospect of allowing a market choice for consumers with regard to privacy. It also has potential for cost reduction, as many a type 1 person could be transformed

> into a type 2 person when costs were made explicit. (Balderston, Carman, and Hoggatt 1977, 1118–19)

Finally, the detailed study prepared by Baxter, Cootner, and Scott (1977) entitled *Retail Banking in the Electronic Age* should be mentioned. Subtitled *The Law and Economics of Electronic Funds Transfer,* the study confirms the argument presented in chapter 1 about the inherent value of the informational commodity for secondary uses:

> Likewise, it is probable that *retailers will generate a wide range of management information from their side of EFTS operations* [emphasis added]. This may be a major potential benefit of some types of EFT systems, aiding retailers in their purchasing and marketing decisions, but it does not seem worth exploring from a privacy standpoint. However, there is a related possible use of customer file information that is worth mention. If a POS system affords sufficient purchase detail, a bank or a retailer will be able to determine who among its customers has bought a given product or type of product and what some of his or her socioeconomic characteristics are. Analyzing large groups of customers can be an efficient aid to market research. (164)

Like the texts mentioned earlier, the Baxter, Cootner, and Scott study suggests that the value of privacy will have to be reconsidered on a more economic basis:

> A more balanced view would regard privacy as a value or "good," which is prized highly by many persons and desired to some extent by almost all. Like any other good, it costs something to produce, and the amount of that cost varies in different circumstances. It is not a homogeneous good; both the demand and the cost will depend on the exact type of information in question, precisely who is being given access to it and for what purpose. (162)

The record of North American privacy policy suggests that controls established in early cable statutes, which appear to nullify major aspects of the informational commodity, will not last long and do not represent a permanent trend. Undermining this trend while still proposing to safeguard privacy will require some ingenuity. A particularly keen solution is hinted at in the commentaries of Westin, Tydeman et al., Flaherty, and Baxter, Cootner, and Scott, as well as in the *Science* article. Attaching an economic value to privacy forces this interest to compete and defend itself in the market place. This effectively shifts the grounds of the privacy debate from a discussion of social values and priorities to a crass consideration (calculation) of individual economic interest. As if this were not sufficient, the degradation of the privacy interest is then couched in a rhetoric of freedom of choice. When considering future trends in privacy policy, this trend toward privacy at a cost will predominate. It is certainly more consistent with the North American tradition of industry legitimation disguised as a defense of rights. When privacy policy is enshrined at the federal level, there is every reason to believe that it will embody the sorts of compromises described above.

Although privacy policy will constitute the primary legal-regulatory constraint on uses of the informational commodity, the discourse on privacy itself is not an adequate theorization of social control in home networking. A major shortcoming of privacy discourse is the manner in which it restricts an analysis of social control and computerized files to the problem of abuse and the individual. It systematically excludes from the field of discourse a consideration of the structural necessity of these practices for the maintenance of modern bureaucratic organizations. The next section examines one critical reformulation (Rule's 1973 analysis of public surveillance systems) of what is at stake in the privacy debate. This reconceptualization of data surveillance and privacy is then applied to an assessment of the informational commodity. Although Rule's work represents an improvement over formulations of the privacy problem implicit to privacy policy, his work does not represent a comprehensive theorization of social control in home networking. In the next chapter, a more commensurate approach is put forward, based on the concept of social management.

3.3 Reformulating the Privacy Issue

There is an alternative to the acknowledged version of what is occurring in the discourse on privacy protection. As Rule (1973) has succinctly stated, the central problem is that of obtaining "compliance from a mass clientele in a large-scale social setting" (28). The social givens in Rule's formula are the "agencies" seeking compliance, their "mass clienteles," and the "large-scale social setting" which is his characterization of the social order. Rule postulates a dialectical relationship between the development and the perfection of more powerful techniques for controlling mass clienteles and extending social scale. Developing better mechanisms of control is a condition for extending social scale; extending social scale demands perfection of new techniques of control (301).

Rule's analysis is notable for its incisive, economical characterization of the principles of what he terms "public surveillance." Nevertheless, the components of his analysis can be reformulated in order to stress their significance more precisely. The social units interacting in the equation are large: public- and private-sector agencies which are organized hierarchically and bureaucratically, with highly mobile and anonymous mass clienteles. The point or focus of control for the management of mass clienteles is not, however, the "mass" itself, but the individual client. Whereas the instrument of control is personal documentation, the mechanism of control is the ability to link a client to his past through this documentation and to use it as basis for decision making. The fact that control hinges upon this particular mechanism is a function of the very nature of the agency/mass clientele relationship, characterized as it is by bureaucracy at one end and anonymity and mobility at the other. This characterization of the control relationship is important because it acknowledges that the first objective of these activities is not to control individuals but to manage mass clienteles, which is itself a structural feature of the social order proper to western industrial economies.

Although agencies may strive for total surveillance as a way of managing mass clienteles, there are inherent economic, technical, and institutional constraints which work to limit such a capability. Rule (38–40) has identified four such limitations on the perfection of systems of total surveillance. The first concerns constraints on both the size of the clientele and the amount of information a system can reasonably cope with and use in its decisions on individual clients. A second constraint concerns the extent to which data files are centralized. This will determine the degree to which disparate client information, generated in different contexts, may be integrated and applied to decision making. A third area in which systems fall short of total surveillance concerns the speed of information flow and decision making. Finally, systems of surveillance are limited by the number of points of contact they are able to establish with their clienteles.

The introduction of home networking represents a major development in the growth of systems of public surveillance. Its impact on techniques of public surveillance will be felt most strongly as these systems increase points of contact between client and agency, because there is no practical limit to the types of information which will circulate through interactive systems. In chapter 1 I noted that a marketable home networking technology will have to offer a package of diverse services. This means not only that multiple services will be channeled through interactive systems but also that a large number of industries, correspondingly diverse in the scope of their activities, must join together in order to render home networking operational.

Virtually any firm or organization which the consumer currently has dealings with will qualify as a participant in home networking. It is technologically feasible for most, if not all, communication between agency and individual to be mediated through interactive systems. This will affect points of contact because it will increase the number of clients interacting with the firm through a common, uniform network. Because communication will be initiated in digital form and not require further encoding by institutional personnel, information will be susceptible to cost-effective, automated processing. The convenience of these systems, due to their installation in the home, will foster more frequent use of services. A parallel development in the informational commodity will be the production of more detailed information on the consumer's activities. Depending on where this information is stored (e.g., in the data banks of the system operator or in the files of the service provider), it will be possible to combine information from different sources to form consumer profiles. Home networking will also reduce time lapses in encounters between client and agency. Where necessary, surveillance will take place in "real time"—however that may be defined by the agency concerned. Home networking will provide the perfect technological solution to the problem of augmenting points of contact in systems of public surveillance. Moreover, it will do this on a scale and with an efficiency which clearly could not be effected otherwise.

It is important to remember that home networking will be implemented in conjunction with other technologies of electronic exchange, such as point-of-sale terminals. Most likely, procedures for consumer access and identity verification will be standardized for all electronic systems of exchange. In France, experiments with "smart cards" incorporating individualized microprocessors as a defense against illegitimate access are forerunners of a future electronic payments network (Magliana 1984). Although Rule does not treat these new technologies in specific terms (largely because his research was conducted in the 1970s), his analysis anticipates their introduction and assesses their significance, with ominous implications for personal liberties:

> The proliferation of such contact points between system and clientele, coupled with rapid information flow, has important implications for the intensity of scrutiny possible over any one client. A system possessing many such points and rapid intake from them can potentially use them to build a continuous, minute-by-minute account of the activities of any one client—or at least of certain of those activities. (1973, 273)

How important are such sophisticated techniques of electronic surveillance for the maintenance of control—defined, in this case, as compliance with regulations? Rule has observed that, in most cases, even if a very small percentage of individuals disobeyed the regulations of a system, it would break down (288). System stability, therefore, is very tenuous. It depends as much on individuals' perceptions of the legitimacy of authority and the fairness of rules as it does on the actual potential for sanctions to be applied for noncompliance. Because deviant activity is a relatively small fraction of actual client behavior, one may question whether surveillance is necessary at all. To what extent is the cost of monitoring the activities of all clients justifiable in the light of what may actually be saved by apprehending a few "deviants"? Would it not be more cost-effective just to let them go, given the costs of mass surveillance and the fact that most individuals appear to comply regardless of the potential for effective sanctions? Public surveillance systems, however, have not been dismantled but are being enhanced, suggesting that client acceptance of both rules and the authority standing behind them is precarious indeed.

Although compliance with rules is related to their internalization, which, in turn, engenders forms of conditioned behavior, this internalization is clearly related to a subject's awareness of an apparatus of surveillance. Surveillance in itself does not ensure compliance, but an awareness of its presence clearly does so by subtly encouraging the individual to internalize the rules. Ultimately, the importance of home networking for public surveillance systems will not be a function of the actual application of sanctions based on information derived from these systems. Instead, it will be a function of the presence of this capacity in the life of the client, coupled with his awareness of its ever-extending reach.

Home networking will effectively integrate the individual into an expanding network of computer-based communications. In doing so, it will greatly increase both points and frequency of contact between individuals and a host of corporate-sector institutions. Therefore, interactive systems will have an obvious value for agencies involved in public surveillance. Agency management of mass clienteles will be enhanced by home networking at two levels. First, these systems of surveillance will improve decision making about individual clients by contributing to the quantity and subtlety of the information which may be gathered and used to form conclusions. Second, and just as significant, interactive systems will intensify the individual's awareness that his daily activities are a matter of institutional record. This will foster forms of self-compliance because it encourages the individual to internalize the rules. In a system where individuals are aware of a high level of surveillance, it is, paradoxically, no longer essential for agencies to "watch" every client. This may prove to be the most enduring contribution of home networking to systems of public surveillance.

To summarize briefly, I have analyzed the informational commodity at the level of privacy policy and at a structural level in which personal documentation is a key to systems of public surveillance. At the level of privacy policy, existing legislation is inadequate to protect privacy in home networking. The trend in future legislation will be toward a weak defense of the privacy interest as threatened by interactive systems. A major development will be a transvaluation of the privacy interest which will oblige it to compete in the economic sphere with other benefits and interests. Rule's reformulation of privacy and personal files is significant because it signals the structural necessity of personal documentation as a basis for managing mass clienteles in modern social systems. Rule argues (convincingly, I believe) that the privacy problem is not one of abuse but of the structural necessity of personal documentation as a basis for public surveillance in modern societies. Rule counters the position that the problem is one of achieving balance and avoiding abuse by contending that, in certain cases, practices of public surveillance may be objectionable in and of themselves. He states:

> My position is simply that these practices are sufficiently undesirable in themselves and in certain of their potential consequences that we would do well to curtail them wherever possible. This judgement applies even in cases where systems operate with justice and on behalf of highly desirable purposes. Indeed the development of popularly-supported "just" systems of surveillance strikes me as more worrying in the long run than those against which popular sentiment is capable of being aroused. (350)

Interactive systems will significantly affect personal privacy as the informational commodity augments the quantity of personal information contained in the data files of corporate-sector institutions. This, in turn, will foster attempts to control this information through legislation or self-regulation. There can be no doubt that protecting privacy in emerging interactive systems will be a major public concern in the years to come. I believe that despite its limitations and historical failures, privacy policy represents an important defense against corporate-sector intrusions into the private lives of citizens. However, my present concern is with analyzing the broad social significance of home networking and cannot be restricted to a limited consideration of social control as defined in privacy policy.

The next chapter first considers the work of several theorists who have analyzed critically the importance of computer technology for society. This serves as a historical and critical backdrop for what will then follow, as the social significance of home networking is reconsidered from a perspective of social control as social management.

4

Social Management and Home Networking

4.1 Introduction

Home networking is not the result of fundamentally new technology. A frequent rhetorical claim is that a new merging of computers and communications is at the root of interactive systems for the home. In fact, nothing could be further from the truth. Computers were being linked with communications to form on-line systems as early as the 1950s ("The Computer Age: Automating the Boss' Office," *Business Week,* 7 April 1956, 59). Time-sharing, the genesis of the computer utility concept, was similarly inaugurated in the early 1960s (McCarthy 1968). The sophistication of these networks has increased steadily over the last two decades, along with organizational efforts to harness the power of the computer. What is new is the inclusion of the home in these expanding corporate networks. What I call "social management applications of the informational commodity" are essentially attempts to increase the efficiency and extend the reach of corporate institutions by applying this information as input to a host of management techniques designed to assure growth and profitability in large-scale private-sector corporations, as well as to assure the longevity of agencies in the public sector.

In the first sections of this chapter, the social history of computer technology is considered in three contexts relevant to an understanding of social control and home networking: (1) a critique of computer technology and social control in the *labor process*; (2) a critique of the "intellectual" correlative of the material component of computer technology, *systems analysis,* and its contribution to management control of organizational complexity in the postwar period; and (3) a critique of the social theory of *postindustrialism,* created for systems analysis as a way of justifying its existence and obscuring its association with corporate-sector interests. Analyzing computer technology in these three contexts serves as the critical backdrop for understanding social control as social management.

4.2 Social Control and the Labor Process

The historical starting point for a consideration of social control and computer technology is the application of this technology to industrial production. It was in the productive sphere that computer technology was first applied to problems of social control. Under the guise of electronic automation and new techniques of scientific management, the computer was instrumental in efforts to rationalize production by advancing management control of the labor process. In this context, social control is synonymous with efforts to control the resistant and at times rebellious human factor of production. To examine the application of computer technology to the rationalization of the labor process, the work of Braverman (1974) and Noble (1977) needs to be considered.

Braverman's work on the degradation of labor under the auspices of capitalist production is well known. My discussion of his work is limited to his critique of management science in the monopoly capital period. At the center of Braverman's thesis concerning the degradation of labor is his analysis and critique of Taylorism. Although his work was discredited very early by both the labor movement and the engineering profession, Braverman argues convincingly that Taylorism was not an aberration, an eccentricity limited to the activities of a single individual. Taylor's methods may have been excessive, but the principles underlying them were consistent with historical trends in the rationalization of the labor process. Braverman identifies three of these: "the *dissociation of the labor process from the skills of the workers*" (113); "the *separation of conception from execution*" (114); and "the *use of this monopoly of knowledge to control each step of the labor process and its mode of execution*" (119).

Within Braverman's framework, management rationalizes production by appropriating the worker's knowledge of his craft. Management science becomes the instrument of this systematic appropriation whereby worker knowledge is transferred into the hands of management. As further gains are made in the separation of conception from execution, a whole infrastructure of "non-productive" labor emerges to replicate on paper the production process of execution (239). According to Braverman, as office staffs burgeon, office labor itself becomes the object of rationalization. A major thesis of Braverman's work is that the latest wave of office automation is simply an attempt to apply principles established in shop-floor rationalization to the office.

In the same tradition but with a different focus is Noble's (1977) seminal work, *America By Design: Science, Technology, and the Rise of Corporate Capitalism.* Noble has analyzed the emergence of modern management by looking at engineers and engineering. The objective of this analysis is a critical reappraisal of the engineer's role in the evolution of modern management, coupled with an assessment of its implications for worker autonomy and the longevity of the modern firm. As a basis for this critique, Noble examines the vital contribution of engineers and, to an equal extent, that of the discipline of engineering, to the formation of management science. Initially, the work of engineers was structurally bound up with the demands of machine production in the industrial era. In historical terms, engineering has been the crucial nexus where capital meets science as a fundamental precondition of machine production. Noble has described the historical allegiance of the engineer to capital in the following terms:

> From the outset, therefore, the engineer was at the service of capital, and, not surprisingly, its laws were to him as natural as the laws of science. If some political economists drew a distinction between technology and capitalism, that distinction collapsed in the person of the engineer and his work, engineering. Even in his strictly technical work the engineer brought to his task the spirit of the capitalist. His design of machinery, for example, was guided as much by the *capitalist need to minimize both the cost and the autonomy of skilled labor as by the desire to harness most efficiently the potentials of matter and energy.* The technical and capitalist aspects of the engineer's work were reverse sides of the same coin, modern technology. (34)

As the twentieth century progressed, the engineer made an ever-increasing contribution to the modern firm, particularly in the science-based industries, where the engineer's efforts were extended to include not only the rationalization of production but also that of "modern technology itself" (55). This was accomplished through "the standardization of scientific terminology and methods, the regulation of patent procedures, the efficient organization of scientific research, and the systematic production of technical manpower" (55).

The engineer's expertise in the scientific/technological dimensions of production was applied to the rationalization of the labor process in all its dimensions, including the human factor. Noble has observed that the necessity of extending control in this manner was understood very early by Charles Babbage, a pioneer of the computing machine:

> Babbage understood that profitable enterprise presupposed that the capitalist would dissociate labor from the productive process and reintegrate the two on his own terms, in such a way as to maximize the surplus value produced by the labor force in the productive process. . . . Babbage realized, however, that while the actual design of machinery contributed to this transformation indirectly, the capitalist had also to undertake to bring

> it about directly through the *deliberate engineering of the workplace and the work activity of labor.* (259–60)

Moreover, as investment in machinery increased as a proportion of the total allocation of capital, it became increasingly necessary to operate this machinery at maximum capacity. This generated further pressures to control the unruly human factor of production.

The "deliberate engineering of the workplace" implied not only the expropriation of craft knowledge and its transfer into the hands of management, but also the eliciting of cooperation on the part of workers. Efforts to engineer the cooperation of workers were subdivided into *industrial relations* and *personnel management.* Industrial relations sought cooperation by improving "the lot of workers"; personnel management was somewhat more sophisticated in its approach, because it combined psychology and sociology within a framework of human engineering. By the 1960s, industrial relations was combined with and subsumed by personnel management to become *human relations.* The purpose of human relations—as analyzed by Noble—is the transformation "of the stubborn 'human factor' of production into an efficient, adjusted part of the corporate mechanism" (286). In effect, its objective is to ensure that individual goals conform with the established goals of the corporation. As modern manager, the engineer shifted his focus from "the engineering of things to the engineering of people" (264).

In very schematic form, these are the arguments presented by Braverman and Noble. The point of departure for both studies is the Marxist critique of antagonistic relations in production under capital. They then apply this critique to structural changes in the evolution of modern production, demonstrating that the insights of Marx are still applicable to the study of relations in production in the monopoly capital period. There is little room for doubt in either work about the significance of trends in the labor process. Both studies reveal that production in the period of monopoly capital is characterized by a loss of worker control and autonomy. This is effected through a systematic expropriation from the worker of his knowledge of production, often culminating in the structural transformation of living labor into dead labor (machinery) under the guise of automation. A structural correlative to this is the necessity of engineering worker cooperation and fidelity to corporate goals. This becomes particularly problematic (and necessary) to the firm in the monopoly period as the movement to rationalize all aspects of the labor process increases the alienation of labor.

Of particular significance here is the relationship between technology and social control in the labor process. Stated more succinctly, this is the problem of the historical importance of technology for management control of the labor force. The analyses of Braverman and Noble suggest that technology has had an impact on social control at two levels.

At the first level, technology is identified with the automation process, understood as the operational embodiment of Braverman's three principles of rationalization. For Noble, it is through the work of the engineer that the production process is studied, broken down, reorganized, and transformed into machinery. Technology at this level signifies machinery, whose implantation is underscored by a progressive transfer to management of knowledge about production. Control is effected by replacing human labor with machinery. The principle underlying control, in this case, is one based on the elimination of the human factor in production.

However, where labor cannot be eliminated, a second level of control is required. Here, technology is called upon to assist in efforts to discipline the labor force. Noble identifies these practices in their contemporary form with human relations activities. Although the role of technology is self-evident in the case of automation, its contribution is less obvious in that of social control in the context of industrial discipline. For the most part, this second level of control is synonymous with strategies developed for worker surveillance during the first industrial revolution (i.e., the factory system).

Clawson (1980) has analyzed the relationship between technology and the rise of the factory system of production. Building his argument on the case made by Marglin (1974), Clawson asserts that the grouping of workers together under one roof was not necessitated by the need for a central power source for machinery, but by problems of worker control. The "putting out" system was not satisfactory to the capitalist because he was helpless to control the worker's hours (labor would work only long enough to earn a subsistence wage); moreover, the capitalist was frequently the victim of embezzlement. Recapitulating many of Marglin's arguments, Clawson states:

> Factories or central workshops with large numbers of workers concentrated in a limited area under the supervision and control of a capitalist or overseer have substantial advantages (from the capitalist's point of view) over the putting out system, even if there is no difference in the technology employed. In the putting out system workers could decide for themselves how many hours a day and how many days a week to work. The result was that workers chose to have a lot of leisure; nevertheless, they were able to earn a subsistence wage. (44)

According to Clawson, it was not technology which determined the factory system, but the need to augment control through more effective surveillance of worker activity. Grouping workers together was not, however, adequate to the task. An additional element was required. This came under the guise of technology. Clawson argues that technology alone did not determine what has come to be known as the factory system for organizing the labor process. Instead, it was technology adjusted to an agenda of social control in the factory that determined this development. In effect, technology permitted a reorganization of the labor process which was more conducive to management control of labor.

Combining the analyses of Braverman and Noble with that of Clawson, it becomes clear that industrial technology has been a significant factor in perfecting techniques of social control in the factory, and, more recently, in the office. Essentially, social control may be viewed as the other side of the degradation argument proposed by Braverman, where the focus is not so much on skills as on problems of organizational control in the capitalist enterprise—particularly in the monopoly period, as the size of organizations has increased considerably.

It is now necessary to elaborate in more precise terms what contribution computer technology has made to social control in the work place during the postwar period.

The response of neo-Marxists of the Braverman school has been that the introduction of computer technology has followed the same pattern as the introduction of previous industrial technologies. In terms of social control, it may be inferred that computerized automation augments control by eliminating the human factor of production from sectors which have remained beyond the grasp of automation. But computer technology has also fostered new and more sophisticated techniques for electronic monitoring of worker activity. For people working with video display terminals, this has meant that management can now monitor such things as keystrokes per minute, frequency of clerical errors, time required to perform tasks, and time spent away from the terminal. (The documentation on this is growing as these practices become more widespread. Mickelwright's 1982 work is an example of union concern about these tendencies in Canada).

Somewhat related to the monitoring of work activity is the processing of personnel files. Electronic monitoring may be viewed as a special type of input to a more comprehensive and fundamental mechanism of control, which is the creation of documentation on workers as a basis for control and decision making. A key to personnel management is the ability to gather data on individual employees and apply it to judgments about them. In the modern firm, the process begins with the application form. Often involving a worker's approval for the release of personal information held by other institutions, it involves extensive testing designed to determine his suitability for a particular job and is updated as performance is monitored and evaluated throughout the period of his employment. Computer technology is an important component of data processing activities in the personnel area, particularly in very large firms which must evaluate the activities of hundreds, if not thousands, of employees. These uses of computer technology are in many ways similar to practices of public surveillance, with the important qualification that they are not "public"; they apply to data surveillance of individuals associated internally with the firm. The same principle, however, underlies both types of data surveillance. Decision making is based on documentation linking the individual to his past. But it would be misleading to suggest that the computer initiated these practices. The principles underlying data surveillance clearly preceded the computer, and in some cases run parallel to its use, as in manual file systems. In general, the impact of the computer has been to make existing practices more cost-effective.

To summarize briefly, I have argued that the use of computer technology in the postwar period has been consistent with historical trends in the implementation of industrial technology. The computer has been instrumental in attempts to achieve higher levels of management control over the labor process. It has been an agent of social control in the labor process at two levels: as a technology fostering higher levels of automation and as a technology which presents management with new capabilities for disciplining labor through more extensive and effective surveillance.

In addition to this, the computer has contributed to forms of indirect control associated with personnel management and human relations activities. At this level there is a certain isomorphism between practices of public surveillance as described by Rule and the treatment of employee files in corporate organizations. One may argue that the rationalization of the labor process based on techniques of data surveillance has a functional correlative in techniques of public surveillance directed at the environment external to the firm (i.e., the handling of its clients). Although the first is concerned with internal conditions of production and the second is directed at problems in the external sphere of consumption, the mechanisms available are essentially similar techniques of data surveillance, based in part on the processing power of the computer and in part on methods of behavioral and sociological research which frequently rely on the computer as well.

Home networking will effectively challenge the boundaries of what is external or internal to the firm, making the client more akin to the worker, at least from the perspective of corporate intelligence about him. It is, of course, dangerous to insist too strongly on this point because of the difference between spheres of production and consumption and the particular dynamics proper to each. However, the same engine of rationalization has been applied in both cases, involving similar methods and technologies, equivalent objectives, and comparable implications for social control.

4.3 The Critique of Systems Analysis

Braverman's analysis of technology in the period of monopoly capital is predicated on the problem of understanding transformations in the labor process from the perspective of labor. This approach requires that he be less concerned with describing specific changes in the science of management during this period than with analyzing the impact they have had upon labor. As a consequence, his work does not consider in any detail the importance of computer technology to problems of management control and decision making in expanding corporate organizations.

Along with successes in automation and the control of labor, and in part because of this, the size and complexity of organizations have increased. Braverman's analysis of this phenomenon concentrates primarily on how the office infrastructure created to deal with this complexity is itself subject to rationalization efforts. Noble's analysis focuses on how management attempts to deal with personnel administration problems in large organizations, primarily through human relations activities. Although both works touch upon the relationship between modern management and the control of large-scale organizations, this problem is not central to their arguments.

Both authors imply that the basis for modern management control is a function of its ability to gather information and develop sophisticated techniques for manipulating and interpreting it, and to then concentrate this intelligence in its own hands. For Braverman and Noble, this is essentially knowledge concerning machine production and the management of labor. Computer technology has helped to create this monopoly of knowledge at two levels: at the machine level, as the technological component of advanced forms of automation, and at the organizational level, as a device which has facilitated the management of the labor force in large-scale organizations. Both dimensions of the application of computer technology concern the internal organization of the firm. A third dimension of computer implementation, somewhat related to the second one, concerns the work activity of "non-productive" labor and problems of decision making and control in large-scale hierarchical organizations. This has often been referred to as the management of social and organizational "complexity."

The principle underlying the use of the computer in this context is similar to that described by Noble and Braverman—control on the basis of knowledge, fostered and supported by the computational and data processing power of the computer. When studied in this context, the issue is not computer technology and the degradation of labor, but the importance of computer technology to management science during a period of growth and expansion among corporate-sector institutions.

In the postwar period, systems analysis has served as the strategic juncture where advances in electronic (computer) technology have been adapted to meet complex problems of management control. In systems analysis, the rationale for the existence of these practices is often defined as the need to secure control of social and organizational units which themselves are evolving at an accelerating pace in the direction of greater complexity. According to this argument, it is the natural and irreversible evolution of social institutions into larger and more extended organizational structures that has compelled the creation of systems techniques. A key component of this argument has been the postulation of a hastened pace of generalized social change in the environment external to the firm. It is contended that large organizations are ill equipped to adapt their decision-making procedures to the realities of social change. Electronic technology in both its dimensions, data processing and communications, is viewed as a tool for accelerating decision making, and, therefore, for keeping in tune with social change. In this section, a number of these arguments are challenged.

The military origins of systems analysis in operations research and in the development of radar during the Second World War are well known (Ackoff and Rivett 1963, 6–7). The transfer of these techniques to the civilian sector in the 1950s and 1960s is similarly a matter of public record. Somewhat more problematic, however, is the question of determining what defines systems analysis as a unitary, systematic approach to management problems. The essential characteristics of systems analysis were outlined for management science by Ackoff and Rivett in their 1963 volume, *A Manager's Guide to Operations Research.* According to these authors, the essential characteristics of operations research include "(1) systems orientation, (2) the use of interdisciplinary (or mixed) teams, and (3) the adaptation of scientific method" (10). One might hasten to add a very heavy reliance on the computer whenever the "systems orientation" prescribed the simulation and modeling of complex structures which could be effected only with the aid of the computer. In terms of the research presented here, attempts at self-explanation and self-justification by the field are less important than the social history of these techniques in the context of their postwar application. As a way of framing the social significance of systems analysis, I will consider briefly the arguments of three authors who have analyzed this phenomenon: Weizenbaum (1976), Hoos (1983), and Mowshowitz (1976).

Weizenbaum, in a chapter entitled "On Tools," challenges the rhetoric of systems analysis when it claims that systems research arrived "just in time" to save military and social service institutions which were approaching the limits of their "internal speed." He accepts the claim that computer technology arrived in the nick of time but questions what precisely it was that the computer and systems research preserved. In Wiezenbaum's critique, computer technology did not save society and its institutions from imminent collapse; it merely preserved the status quo. More important, in doing so it impeded critical public reflection on the trend toward larger, more hierarchical, and more bureaucratic corporate institutions. Weizenbaum states:

> Yes, the computer did arrive "just in time." But in time for what? In time to save—and save very nearly intact, indeed, to entrench and stabilize—social and political structures that otherwise might have been either radically renovated or allowed to totter under the demands that were sure to be made on them. The computer, then, was used to conserve America's social and political institutions. It buttressed them and immunized them, at least temporarily, against enormous pressures for change. (31)

Weizenbaum's argument is provocative, because it challenges the accepted perception of computer technology as a force behind radical social change. According to his critique, computer technology is a *conserver* technology when viewed at a macro level in which entrenched social and political structures seek to maintain their authority and control. Systems analysis, when interpreted in this light, is a powerful instrument for the *preservation* of corporate-sector hegemony.

In a very detailed and well-documented study, Hoos has analyzed the record of systems analysis in its public-policy applications. Whereas Weizenbaum implies that systems technology has been an effective management tool, enabling management to maintain rational control over expanding corporate bureaucracies, Hoos contends that the systems approach, when applied to public policy, is essentially irrational. Its real importance has been to legitimatize forms of elitist technocratic decision making at the expense of public forums of democratic debate, particularly when the subject of public concern has been the assessment of new technology. Hoos summarizes her conclusions in the following terms:

> Without belaboring points made in earlier chapters, we might do well to keep in mind some of the salient features of systems methodology in theory and practice: (a) it is intolerant of and even antagonistic to unquantifiable variables; (b) it contains a bias against social and human factors; (c) its ethic is that of the marketplace; (d) it is driven by technical optimism. (1983, 248)

Hoos demonstrates that systems methods have succeeded in monopolizing policy discourse to the exclusion of less technically optimistic, more qualitative, and value-oriented approaches. The importance of systems analysis to policy research has been to smooth the passage to ever-higher levels of technologically supported rational management of the social order; this is done by creating a benevolent public face for technological rationalization. It goes without saying that this rationalization process is fostered by corporate sector-organizations in accordance with their long-term interests of growth and expansion. In this context, systems analysis functions as the necessary ideological correlative to corporate efforts to maintain control through forms of technological rationalization which may be objectionable or harmful to society, and therefore the subject of controversy.

Mowshowitz's work, aptly subtitled *Information Processing in Human Affairs*, directly challenges the "myth" of social and organizational complexity. The growth of computer technology and the techniques of systems analysis associated with it are functionally related to elitist decision making and the centralization of power in government and private industry. The myth of complexity obscures this relationship.

Mowshowitz demonstrates that the complexity argument is misleading in several important respects. Although the term *complexity* is common currency in systems analysis and management science—as well as in a number of other disciplines, including (to a lesser degree) sociological, political, and economic discourses—it does not stand for a "unitary concept" (1976, 232). Mathematical measures of complexity developed for the description of *disorganized* physical systems are not appropriate for describing structures of *organized* complexity as exemplified in social systems. Complexity is not a measure or characterization of the objective state of things (or "systems") as they exist in reality. Instead, it is more properly an index of epistemological uncertainty. It is an indicator of the uncertainty of the social scientist when confronted with phenomena whose internal structure and functioning he is unable to decipher.

The notion of complexity is far from illuminating when applied to social structures. More significant, however, is the question of what has been occluded in management science and systems discourse under a rubric of complexity. Mowshowitz shows that the type of complexity described in management discourse as a justification for systems analysis is specific to modern bureaucratic organizations, structured hierarchically with control and decision making concentrated at the top. Explicitly, he challenges the underlying assumption of Simon's theory of organizations that hierarchy is a universal and inescapable feature of organizational structure: "Hierarchical structure in large organizations is not an intrinsic feature of social evolution. It is a response to a particular organizational goal, and is cultivated in order to achieve the disciplined effort of a large number of different individuals in the pursuit of that unitary goal" (79). The management of complexity, the self-proclaimed objective of systems analysis, reveals itself as a particular problem of control proper to the management of modern bureaucracies.

In systems discourse, complexity is both a characterization of the problem facing modern corporations and a justification for the systems approach as a way of coping with it. Contrary to the claims of management science, Mowshowitz demonstrates that complexity is not a problem endemic to a particular stage in the "natural" evolution of society and its institutions. The complexity rationale conceals the reality of management attempts to maintain centralized control in organizations oriented toward an expansion of their reach and influence while maintaining (internally) hierarchical structures of authority and decision making (227–47).

By combining the critiques of Weizenbaum, Hoos, and Mowshowitz, a picture of the social significance of systems analysis emerges which radically contradicts the discipline's own portrait of its function and value for society. These authors demonstrate that systems analysis is essentially a conserver technology of status-quo political and economic structures in the postwar period. In its public policy applications it has served an ideological function, monopolizing the discourse on social futures, deflecting objections to technological rationalization, and transforming political issues into technical problems to be treated by specialists. In its management applications, it has worked to preserve hierarchical bureaucracies with centralized decision making against pressures to decentralize. The myth of complexity, as maintained in systems discourse, legitimates modern hierarchical bureaucracies. Moreover, it suggests that they are inevitable and permanent—when, in fact, these structures are contingent upon particular organizational goals proper to capitalist enterprise and upon the demands of a disciplined work force working toward unitary corporate goals.

4.4 Systems Theory and Social Theory

The problem of social complexity has implications and reverberations in social theory which go beyond a consideration of organizations. To the extent that systems analysis presents itself as a technique for managing social complexity, it necessarily incorporates assumptions about the larger social order which surrounds its efforts. These assumptions are largely implicit in systems theory and are not usually elaborated on in explicit form. For the most part, however, the theory of society implicit in the systems approach corresponds to the theory of the postindustrial or information society as described by a number of social theorists, particularly by Daniel Bell. It is important to consider systems analysis in the context of its relationship to the social theory that Bell created for it, the theory of postindustrialism.

In its most general form, the theory of the postindustrial society may be understood as an amalgam of concepts, propositions, and formulas embodied in a number of texts published since the mid-1960s. Its central thesis forms a basis for argumentation in several diverse works, including, among others, those of Etzioni (1968), Drucker (1971), Brzezinski (1970), and Bell (1973). This thesis may be summarized in four propositions: (1) the Western industrial nations are in a period of historical transformation of a radical nature; (2) this transformation is such that the future society will supersede or transcend previous industrial societies; (3) the changes which mark this transformation are already visible in current social trends (social indicators); and (4) for the most part, the agent of this transformation is the postwar legacy of technology—both material (electronics, computers, communications) and intellectual (cybernetic theory, systems theory, operations research, game theory).

In the works in which postindustrial theory has appeared, it has served different functions. In most cases, it is a starting point for a polemic which attempts to (a) interpret the meaning of this transformation for society and to (b) identify its implications in terms of future problems for society. The diversity of interpretations of this transformation is suggested in the different labels ascribed to it, such as the "post-modern period" (Etzioni 1968, 9), the "knowledge society" (Drucker 1971, 261), the " 'technetronic' society" (Brzezinski 1970, 9), and the "post-industrial society" (Bell 1973). In the examples cited above, the thesis is usually implied or assumed as a basis for further argumentation; the fact of a radical transformation is not at issue, only its interpretation.

It is primarily in Bell's work that the thesis itself is subject to considered analysis. Bell directs particular attention to an exploration and validation of the role that the material and intellectual (i.e., systems) technologies associated with the computer have played in provoking and guiding social transformation.

In the introduction to his work, Bell describes the "concept" of the postindustrial society as a "large generalization" containing five dimensions:

1. Economic sector: the change from a goods-producing to a service economy;
2. Occupational distribution: the pre-eminence of the professional and technical class;
3. Axial principle: the centrality of theoretical knowledge as the source of innovation and of policy formation for the society;
4. Future orientation: the control of technology and technological assessment;
5. Decision-making: the creation of a new "intellectual technology." (1973, 14)

Bell's dimensions are not all of the same order or status. Some are more straightforward and less interpretive than others. The growth of the service sector, for example, is not an issue. Statistics demonstrate a shift in U.S. employment distribution to service-related occupations at some point in the 1960s. Closely related to this shift is Bell's second dimension, the rise of a technical and professional class. Using statistics related to employment distribution and education, Bell argues that the working population is increasingly better educated and engaged in white-collar jobs; moreover, he argues that within this group professionals, engineers, and scientists are becoming the pivotal class (17, 212–61). Both of these trends serve as evidence to support the claim concerning changes in the social order.

The key dimension in Bell's concept of the postindustrial society is his axial principle, "the centrality of theoretical knowledge as the source of innovation and policy formation for society" (14). The notion of axial principles is central to Bell's method of social forecasting. The creation of a service economy and the rise of a professional and technical class are "extrapolated social trends" (18). The centrality of theoretical knowledge, however, is the "defining characteristic of a social system" (20). Bell acknowledges that knowledge is important to the functioning of any society. His argument, however, depends on the thesis that in the postindustrial society there is (1) change in the "character" of knowledge and (2) this change is the defining characteristic of a fundamentally different form of social organization, as revealed in the role of theoretical knowledge as a force behind innovation and a component of decision making. He states:

> What is distinctive about the post-industrial society is the change in the character of knowledge itself. What has become decisive for the organization of decisions and the direction of change is the centrality of *theoretical* knowledge—the primacy of theory over empiricism and the codification of knowledge into abstract systems of symbols that, as in any axiomatic system, can be used to illuminate many different and varied areas of experience. (20)

At this point, Bell's third dimension dovetails conveniently with his fifth dimension, "the creation of new intellectual technology" (5). The reference here is to techniques, disciplines, and theories developed since the war in conjunction with the rise of systems analysis. These include information theory, cybernetics, decision theory, game theory, utility theory, and stochastic processes. The material correlative to these "intellectual" technologies is the computer. Bell contends that developments in intellectual technology hold great promise for the "management of organized complexity" (28). Their decisive role in the future society will be in the making of decisions.

Within Bell's theory, this fifth dimension has at least two functions. First, the development of intellectual technology since the war is itself evidence to support the growth of theoretical knowledge, because much of it is grounded in abstract theory. Second, and more important, is the role of intellectual technology in the resolution of society's problems. Bell interprets the significance of intellectual technology as a problem-solving function. In many respects, this is just a variation on the now-familiar "technological fix." It also defies recent experience.[1]

The real agenda of Bell's work is suggested in the fourth dimension, "the future orientation" of postindustrial society. By referring to the planning of technology as a function of a future orientation, Bell underscores the importance of his own field of discourse. In effect, he emphasizes the importance of futures research (forecasting, planning, and policy assessments) to social decision making in the postindustrial era. Moreover, by focusing such attention on technology as the preferred object of planning and control, Bell is able to reify technology, casting it in its familiar role as the agent of social change. In doing

1. As Hoos has pointed out, systems analysis was the backbone of the military's attempt at improved management and decision making in the 1960s. Under the guise of McNamara's PPBS, it was subsequently extended to civilian institutions in government. As a technique it was instrumental in the "victories" of Vietnam. It was so successful at rationalizing management in the defense department that to this day the Pentagon is recognized as one of the most wasteful and inefficient branches of government. In a peculiar twist, Hoos has noted that "*scientific management* has been the device used to rationalize and camoulflage the abuses" (1983, 60).

so he implicitly hides the fact that specific institutional interests, along both political and economic axes, are responsible for decisions to develop and implement technology.

In effect, Bell's exposition is an attempt to create a better society for science and technology. More precisely, it is an attempt to create a good theory of society for a host of postwar intellectual technologies associated with the advance of the computer. It was by no means a coincidence that the emergence of a theory of postindustrial society in the late 1960s coincided with efforts to transfer a systems approach to management from the military to the public sector (i.e., McNamara's "Program Planning Budgeting System)" (358). Bell's ruminations, however, do not stop with the assertion that these new intellectual technologies are useful or even vital to society. According to Bell, they define society itself. They reveal what is different and unique about this new society and its structures of organization when compared to previous societies.

Given the parameters of Bell's scheme, the pivotal question for an analysis of postindustrialism must be: What kind of society is a society defined by the centrality of theoretical knowledge? Because of his background in theoretical sociology, Bell contextualizes this question in terms of macrosociological theory in the tradition of Durkheim, Marx, and Weber. It is with reference to their work that Bell proposes to understand the significance of the new intellectual technologies. Although Bell goes to great lengths to avoid the charge of historicism, his model of the new social order is clearly underscored by evolutionist presuppositions (Goldthorpe 1971). Under a rubric of postindustrialism, Bell proposes that society has evolved along a different historical path from that predicted by Marx. Accordingly, the social order he describes is one in which the crisis of capitalism has been overcome and the many undesirable features of industrialization have been surpassed.

Somewhat more astute than many of his North American counterparts, Bell is prepared to acknowledge the theoretical relevance of Marx's analysis of early capitalist forces and relations of production. Marx's work is only temporarily relevant to Bell's scheme, however, because he quickly suggests that a more appropriate frame of reference would be that of industrialism and postindustrialism, and not that of capitalism. Bell justifies the transition from a Marxist framework to an "industrial–post-industrial" theory of economic relations by arguing that Marx failed to anticipate a number of key "structural" adaptations proper to late capitalism. In his view, these were (1) the development of the credit banking system, (2) the separation of ownership from management, and (3) the expansion of office personnel and white-collar work (1973, 59–60). In effect, Bell argues that these structural changes altered the character of capitalist production to such an extent that the inevitable collapse of capital, as foreseen by Marx, was permanently forestalled.

It would be difficult to deny the sorts of changes described by Bell. It is clear that these later developments in the evolution of capital relativize some aspects of Marx's analysis, particularly those internal contradictions of capital which Marx predicted would spell its demise. But Bell does not end his critique here. Instead, he is prepared to conclude from this that the entire theoretical-critical apparatus constructed by Marx is invalid for a discussion of contemporary society. Moreover, he implies that the capitalist system of production is no longer a relevant object of social analysis. Accordingly, Bell shifts the focus of analysis from the study of social organization under the auspices of capital to a universalizing theory of postindustrial society.

The manner in which Bell sidesteps capitalist production as the economic context for social analysis seriously impoverishes his account of contemporary social organization. It is not within the scope of my discussion to recount in detail the arguments embodied in alternative neo-Marxist analyses of "late" (Mandel 1978) or "monopoly" (Baran and Sweezy 1966; Miliband 1969; Braverman 1974) capital. In many important respects, however, these studies not only challenge Bell's conclusions, but are in themselves superior accounts of the social developments which are the subject of his work. The work of Mandel, a neo-Marxist, is considered here as an alternative economic analysis to Bell's theory of postindustrialism.

In Mandel's work, a reduction in the turnover time of fixed capital is an important feature of, and a central problem for, late capitalist economies. The acceleration of technological innovation lies at the source of this development; the impetus toward technological innovation is itself the product of an ongoing reallocation of capital into research and development fueled by past successes. The net result of these developments "has adverse physical and technical repercussions on the length of life of fixed capital, both because it increases the speed at which machines are used and because it accelerates their obsolescence" (1978, 228). Moreover, this affects circulating capital which, increasingly, must be converted into fixed capital (machinery). Consequently, companies must draw more and more of their circulating capital from financial institutions. Ultimately, capitalist production becomes much more "hazardous" in the late monopoly period than it was in the era of "freely competitive capitalism or 'classical' monopoly capitalism" (228). The necessary response from industry is to carefully plan all aspects of commodity production, marketing, and sales; this is conjoined with government-industry attempts to coordinate the programming of the national economy.

According to Mandel, the ramifications of this for the structure of social organization are clear: the types of control which late capital must exercise over all dimensions of production and distribution are "impossible without a growing regimentation of the economic and social life as a whole" (241). Bell acknowledges this social problem when he notes that one of the central conflicts in "post-industrial" society will be that between citizen demands for increasing "participation" and the growth of "bureaucracy" (1973, 115). Bureaucracy, however, is described as an abstract, autonomous force, devoid of any functional relationship to a specific mode of economic production. It appears as a necessary evil of any industrial society and is problematic only when it gets "out of control."

Mandel's critique is more incisive because it correctly describes the growth of bureaucracy not as a necessary corollary to postindustrial economic expansion, but as the expansion of the net of social control compelled by an ongoing crisis in the mode of economic production of late capitalism. As technological innovation both accelerates the obsolescence of machines and increases the portion of capital tied up in machinery (fixed capital), the pressures on capital to guarantee a return on this investment increase enormously. These pressures manifest themselves as a corporate agenda to forecast, plan, and control all aspects of production and distribution. This agenda translates into the development of specific techniques—many based on the computer—designed to reduce uncertainty and control events in the factory, the office, the marketplace, and society at large; hence Mandel's regimentation of society. Whereas Bell is prepared to see in the social indicators of postindustrialism the potential for continuous economic expansion, Mandel and other radical theorists have viewed current developments from the perspective of an ongoing crisis of late capitalism. It is not the inexorable and autonomous momentum of technological and social progress that propels change and destabilization, but a well-known problem of profit extraction in a production system which is in a perpetual state of crisis.

In a somewhat less ideological vein, Warskett (1981) has reformulated Marxist theory, and particularly the work of Mandel, into a theory of modern "information" economies relevant to policy research. He explores some of the contradictions implied in state efforts to program (coordinate) long-term economic planning in an economy of competing capitals. Warskett's thesis is summarized in the following terms:

> This paper argues that large corporations, as forms of highly centralized and concentrated capital, are responsible for the transformation of the economy in the phase of late capitalism into an information-based economy. Concentrated forms of capital must monitor and control their economic environment, that is, take on the attributes of a cybernetic system in addition to their normal productive activities. (179)

He notes that large corporations must strive for what he terms "cybernetic" control because the price mechanism is no longer valid as an allocator of corporate resources or as a device for determining corporate strategies. Forms of concentrated and centralized capital must rely less on the market system and more on "discretionary intelligence operations" (carried out internally) as ways of effecting long-term strategy. A whole category of nonproductive informational labor emerges. Its function is to reduce uncertainty in the environment affecting the firm. Warskett states:

> Market prices are cybernetically inadequate as signalling devices, as message carriers in an economy dominated by large corporate capital units. Competition still exists, but it is largely of a non-productive variety. At the abstract level this form of competition here is termed cybernetic competition. One concrete manifestation of this competitive form is oligopolistic competition which largely eschews the price mechanism as a system of market signalling. (185)

He continues:

> Much more reliance is placed in non-price competition and hence upon information about *consumer demand elasticities, rivals,* and *potential entrants into the industry* [emphasis added]. In this way the leading modern corporation is induced to undertake the production of information itself or purchase it from *consultants, advertising agencies, market research firms* [emphasis added], etc.

Contrasts between Bell's postindustrial economy and Warskett's information economy of cybernetic competition are striking, particularly regarding their different interpretations of the function of postwar intellectual (i.e., systems) technology. In Bell's postindustrial scheme, the computer and the systems approaches associated with it are vital adjuncts to the economic order. They are "savior" technologies which assure social stability in an economy defined by its commitment to potentially disruptive technological innovation. In Warskett's scheme, these cybernetic technologies are not auxiliary to the economic order; they define its very organization. Warskett's choice of the term "cybernetic" to define the competitive order of late capital suggests the vital role which systems techniques have played in assuring the proper functioning of oligopolistic capitals during this period.

For both Mandel and Warskett, planning—particularly for the long term—is essential both to individual oligopolistic capitals and to the larger economic order which situates and sustains them. The corporate sector usually ascribes benevolent objectives to its planning activities. Chiefly, planning is identified with a more "rational," considered allocation of resources in the future. According to Mandel and Warskett, corporate-sector planning is not so much benevolent or rational as it is inescapable, given the economic order of late capitalism and its trend toward a reduction in the turnover time of fixed capital.[2]

2. It is not my intention to attribute the crisis of late capitalism solely to the problem of a reduction in the turnover time of fixed capital. There are many problems threatening the economic order

In the previous sections, computer technology has been examined from the perspective of a number of existing critiques of computer systems. These critiques demonstrate an important synergism between the practical application of computer technology for purposes of control and the fabrication of a discourse on computers and society designed to occlude this fact. This discursive-ideological component is realized in its most totalizing form in the social theory of postindustrialism. The ultimate import of this synergism is a tightening of the web of social control. The perfection of computer-based techniques for corporate-sector management of the social order is effected concurrently with the obfuscation of corporate power and interest in social discourse on computers. Corporate-sector control is a function of both the application of control techniques derived from the computer and the hegemony of discourse. Systems discourse and the discourse of postindustrialism both work to obscure corporate interests fostering this project to computerize society.

4.5 Social Research for Social Management

A focal point for the preceding analysis of computers and corporate power was the critique of systems analysis. This critique revealed the hegemony of systems discourse in the policy process (Hoos) and challenged assumptions about the social order implicit to systems analysis, showing their link with corporate-sector interests. It is important to remember, however, that systems analysis is not just discourse and ideological subterfuge. Hoos's catalog of "systems" failures to the contrary, systems analysis works. This is not meant to suggest that it works at the level of and according to its self-proclaimed aspirations to effectively simulate social, political, and economic systems (at either a micro or a macro level) in any real sense. But it functions well enough as a tested component of contemporary social research—particularly when it is future-oriented and concerned with the behavior of aggregate social units—to have established its value for the corporate-sector planning and forecasting so vital to the cybernetic economy. The connection between systems analysis and social management (as a form of social control) is best understood from the perspective of social research following the war.

Attempts to reevaluate the objectives of social research in the light of com-

---

of corporate capitalism, but the acceleration of technology turnover is nevertheless a major concern. As elaborated by Mandel, it is the best way of explaining the relationship between material and intellectual technology. It also explains the process of technological rationalization which begins in preproductive activities of science-based research and development and culminates in the public sphere in the form of computer-based systems analysis proposed as a way of guiding and legimating this process. For a more comprehensive assessment of the micromodalities of this crisis, I defer to Edward Herman's listing of the problems facing the corporate power structure (Herman 1981, 248–50).

puter technology and new capabilities for the analysis and management of the social order may be traced to the dawn of the computer era. Lewin ([1947] 1969) presented the case for social management by demanding whether there was anything equivalent in social life to the self-steering mechanisms characteristic of advanced missile systems.[3] He concluded in the affirmative, with the important qualification that these mechanisms were primarily ad hoc and not adequate to the challenges facing society in the future. A major problem, according to Lewin, was the lack of sufficient data for social management: "An efficient steering of social action presupposes that fact-finding methods have to be developed which permit a sufficiently realistic determination of the nature and position of the social goal and of the direction and the amount of locomotion resulting from a given action" (1969, 442). Lewin argued that social scientists would have to accept the challenge of dealing with larger social units and periods of longer duration if they were to meet the demands of social research in the future: "We will have to learn to handle relatively larger units of periods and social bodies without lowering the standards of validity and reliability to which we are accustomed in the psychological recording of the more microscopic units of action and periods of minutes or seconds of activity" (1969, 443). He concluded by extending his challenge to society at large, urging that social management would be a key to the maintenance of democratic government:

> Obviously social management in the various areas of modern society have [*sic*] to face a tremendous task. Its solution presupposes social fact-finding of an unheard of magnitude. It requires basic research about social steering systems. The fear of fascism seems to have driven some people into the greatest kind of misunderstanding which identifies democracy with planlessness. The survival and development of democracy depends not so much on the development of democratic ideals which are widespread and strong. Today, more than ever before, democracy depends upon the development of efficient forms of democratic social management and upon the spreading of the skill in such management to the common man. (1969, 444)

In the late 1960s, a similar conceptualization of social management resurfaced under the rubric of *social engineering*. In an appearance before the U.S. Senate Subcommittee on Intergovernmental Relations, Michael (1967) assessed the impact of new technologies of social engineering for government. His comments are significant in at least two respects. First, the timing of his

3. Kurt Lewin, "Feedback Problems of Social Diagnosis and Action," in *Modern Systems Research for the Behavioral Scientist*, ed. Walter Buckley (Chicago: Aldine, 1969), 442. The original article appeared in 1947 (Kurt Lewin, "Frontiers in Group Dynamics," Part II-B, *Human Relations* 1 (1947): 147–53). Materials cited here are drawn from Buckley (1969).

presentation corresponds to the moment when systems analysis, broadly defined, was beginning to make its appearance in the civilian branches of government and in the private sector, as opposed to its previous military applications. Second, his articulation of the problems facing society and the role of new computer technology in their resolution is representative of the project of social management. Michael proposed the following definition of social engineering for his congressional audience: "By social technology or social engineering, I mean the deliberate application of systematically accumulated knowledge and theory about the nature of man and his institutions for the purpose of influencing the behaviour of man and his institutions" (888). He acknowledged the origins of social engineering in military control systems and signaled a number of precursor projects in the nonmilitary sector:

> This area of technology is already with us in its early and relatively primitive forms. Examples would include the human engineering of weapon systems; the use of systems analysis, and program budgeting and planning systems in the development of long-range economic planning; Keynesian-type economic interventions; the application of behavioral science knowledge to urban design; the research leading to and the evaluation of Operation Head Start; the extensive work in new managerial theory and application, especially as applied to industry. (888)

According to Michael, the emergence of social engineering could be explained by the convergence of the social needs of the welfare state (i.e., "the elimination of poverty, the enhancement of education, the improvement of old age services") and new technological capabilities offered by the computer. Computer technology was crucial to the social scientist because it provided him with capabilities which had previously been beyond his grasp. First, it would give him "the means for combining in complex models as many variables as he needs in order to simulate the behaviour of men and institutions," and second, the social scientist could "for the first time test these models against conditions representing 'real life' " (889). This second condition applied because the data processing power of the computer would make it possible to monitor the state of society in real time. Michael continued: "Thus, the social engineer not only can know the state of society now as represented by these data, but he can use them to test and refine his theoretical models in ways analogous to those used in the development of simulation models for describing the behaviour of rockets, aircraft, etc." (889). In the light of these developments, "deliberate social manipulation" would be required in the future:

> And given the convergence of the powerful physical technologies and social technologies, and given our already enormously complex and huge society, it would seem that deliberate social manipulation will be neces-

> sary if we are to introduce appropriate changes in society at the appropriate time. (889)

According to Michael, the crucial question for Congress and for society would be "*who* is to decide *whom* is to be manipulated and for *what ends*" (889).

Michael accepts a priori the future necessity of social research for social management. In this respect, his remarks reflect the perspective of social researchers and institutions with a vested interest in social management. They are somewhat atypical of this community, however, in another way. Although Michael argues that social engineering is inevitable (an argument which I reject), he does not pretend before his congressional audience (using the language of futures research, for example) that social engineering is inherently democratic or liberating. He argues that the agenda of social engineering is "deliberate social management" (891). Moreover, he speculates on some of the ways in which new technology in general, and social engineering in particular, would affect "cherished values." Concerning problems of privacy and of the programmed economy, he states:

> 2. We will need *increasing amounts of information about people and their activities in order to better plan* [emphasis added] for their growth and development. Obtaining such knowledge is bound to confront us with invasion of privacy issues. What will be the balance between future social planning needs and traditional values and expressions of individual privacy?
> 3. Long-range planning and assignment of social priorities will necessarily conflict with some meanings of the idea of private business freedom. What is to be the basis for trade offs between these needs? (891)

It is important to note that, as a social scientist, Michael does not argue that social engineering should be curtailed; on the contrary, he defends the position that what is needed is more data for social research. Michael's statements imply that the alternative to social engineering is social chaos. Accordingly, the tradeoff between a more regimented society and social engineering is a social necessity rather than a matter of public choice. One may differ very strongly with Michael's assumptions about the unavoidability of social engineering, but his candor about the manipulative aspects of social research for social engineering is uncharacteristic of the field. In particular, his identification of the crucial political question "*Who* is to decide *whom* is to be manipulated and for *what ends*" goes to the heart of the issue of social management.

It is this fusion of social research, computer technology (in both its material and its intellectual dimensions), and corporate economic interest (necessity) which has spelled the birth of computer-based social management in the postwar period. Based on the statements of Michael and Lewin, social management may be viewed as a particular orientation of social research geared to the demands of long-term planning and aggregate social control of the cybernetic economy.

The informational commodity of interactive systems will serve as input to the social management activities so vital to the cybernetic control of the information economy of late capitalism. When this function is grasped, a very different understanding of the social significance of home networking emerges from the one implied by Bell's postindustrial theory. Home networking is no longer a technological gateway to enlightened decision making and participatory democracy in a pluralist society; it is a new source of input to an existing corporate communications infrastructure directed at managing and controlling the social order.

4.6 Social Management and Futures Research

Although different labels have been employed to describe it (social technology, social management, social engineering), the historical project of social research for social management has remained consistent since the advent in the 1940s of electronic cybernetic systems and the first electronic calculating machines. As noted in the discussion of Greenberger's work in chapter 1, section 6, the 1960s project for a computer utility was viewed as tremendously beneficial for social research. Since the 1960s and the political controversies associated with the computer utility concept, social scientists have been more attentive in their use of language—at least when it has been directed at the general public. The manipulative connotations of "social management" and "social engineering" have effected a linguistic change in the discussion of social research for social management. They have been replaced in current parlance with the more positive terms *social planning* and *social forecasting.* Moreover, these activities have been legitimated through their incorporation into the discipline of futures research.

Elsewhere, in a genealogical critique of futures research, I have traced the origins of social forecasting to technological (i.e., weapons) forecasting for the military and the Rand Corporation think tanks established by the U.S. Air Force after the war (Wilson 1982). In the early 1950s, methods of normative and exploratory forecasting designed at Rand for military technological forecasting were applied to problems in the private and public sectors. Optimism subsided, however, when it was discovered that social factors were impinging upon the technology transfer process, undermining the accuracy of technological forecasts. This was suggested in a major forecasting journal, *Technological Forecasting and Social Change.*

The journal *Technological Forecasting* was first published in 1969. In the first volume, a number of articles hinted at the relationship between sociological and technological forecasts (Linstone and Lenz 1970). In the following year, support of this perception was such that the title of the journal was modified to include a reference to social factors: *Technological Forecasting and Social Change,* the title which has survived to this day.

The importance of combining sociological and technological forecasting was underscored in an article by Abt (1970) in the second volume. According to Abt, "social forecasting is the necessary complement to technological forecasting for effective R & D planning" (17). That a consensus among technological forecasters had been reached on this issue was reflected in the alteration of the journal's title to include a reference to social change. In more specific terms, Abt argued that forecasters would have to factor into their calculations what he termed the "*social efficiency*" of technology:

> By forecasting technology we can estimate what will be technically possible in the future, but we cannot from this alone determine what technologies that are possible are likely to be exploited, or to what degree. What technologies are actually exploited will be determined both by their relative technical efficiency and by what might be called their *social efficiency.* The social efficiency of a technology would be a measure of how well it responds to massed public needs and demand (and we must be careful to maintain the distinction between need and demand). These public needs and demands are of course composed of social, cultural, psychological and political factors within economic and political constraints. (17)

My initial research revealed that the discipline of futures research emerged neither as a response to generalized public interest in the future nor because of a spontaneous fascination with the "year 2000" as suggested by apologists such as Bell (see his "Introduction" in Kahn and Wiener 1967, xxv). Futures research, with all its benevolent pretensions and objectives, was merely the public face of a new field oriented (in the tradition of Lewin's remarks) to assessing the future behavior of aggregate social bodies and periods of long duration: social forecasting. Moreover, the mandate for social forecasting is not as broad as its name would suggest. As Abt indicates, the research agenda for social forecasting coincides fairly consistently with the objectives of its progenitor in technological forecasting. The bias of social forecasting since its inception has been toward a better understanding of the social dimensions (implications) of technological change. In the information economy of cybernetic competition, this is the problem of risk reduction in a corporate economy dominated by a progressive decrease in the turnover time of fixed capital.

Incorporation of social forecasting under an umbrella of more generalized futures research obscures the bias of social forecasting toward the objectives of purely technological forecasting and the corresponding corporate-sector interests which finance and promote it. Futures research justifies its existence by proposing that it seeks to analyze the implications of current technology transfer and to sketch alternative futures as a way of keeping the future open to public choice. Numerous critiques of futures research (including my own: Wilson 1982, 1–113) show, on the contrary, that futures research is essentially elitist and motivated by corporate-sector interests (e.g., Goldthorpe 1971; Kumar 1978; Miles 1978; Wynne 1975). According to this point of view, such research effectively closes off the future by restricting public debate on technology to an instrumental consideration of the means toward technological progress, while deflecting attention away from a consideration of social values and goals. It is not within the scope of my analysis to recapitulate this critique in full. It is relevant, however, to note that the rhetoric of futures research (concerning its benevolent goals and social function) also works to cast a favorable light on social forecasting and planning. The negative connotations of social engineering and social management are transformed into future-oriented social research which purports to forecast not in the interests of control or manipulation, but in the benevolent interest of opening the future to human choice.

4.7 Social Management versus Public Surveillance

In the previous chapter, forms of data-based social control were identified with practices of public surveillance. To recapitulate briefly, Rule's (1973) concept of social control as public surveillance is predicated on ensuring compliance with certain rules of behavior (concerning the credit relationship, the conditions for granting services, the enforcement of tax obligations, and so on). The materials presented in this chapter reveal that the dimensions of data-based social control go beyond Rule's identification of it with agency control of mass clienteles. I have characterized these other practices as social management in order to distinguish them from public surveillance activities.

Social management as a form of data-based social control is operated through forms of polling, behavioral research, statistical research, and social forecasting and planning which rely heavily on the computer and database technology but do not apply to the conditions of public surveillance. Although information may originate in the files of individual data subjects, it is not necessary for purposes of social management that it be retained in personally identifiable form. Social management is not directed at assuring the compliance of mass clienteles to a system of rules. It does not seek to apprehend the individual in his life activities but to forecast and simulate the behavior of aggregate social units. As such, it is an effort to control events, to manage the future. Although these practices affect relationships of power among individuals, communities, and society at large vis-à-vis the corporate-state, they are not highly politicized because they do not threaten (or appear to threaten) the foundations of liberal political philosophy regarding the rights of individuals. Nevertheless, they are practices of social control, grounded in the material and intellectual technologies of the computer.

Organizations in both public and private sectors whose raison d'être is the provision of services will normally deal with mass clienteles and problems of compliance necessitating the creation of personal files. But the way they use this information is not restricted to practices of public surveillance. Information extracted from personal files may be designated for secondary use in social management activities like social forecasting and planning which do not require the retention of data in individualized form. The overall objectives and basis for longevity of a service institution will depend on more than the management of its clientele through surveillance of the individual. The same information may be applied to internal analysis of an organization's efficiency and productivity, to forecasting, planning, and simulation efforts, and to the development of market strategies. In the private sector, these additional uses of information gathered from client documentation may be the source of competitive advantage. Within the public sector, this information may be employed to legitimate an agency's very existence (as proof of the demand for services), to justify the extension of services, or to serve as a lever in the context of interagency relations (i.e., an exchange of information profitable to both agencies).

Information extracted from personal files becomes bureaucratic capital to be exchanged and expended according to the mutual advantage and prerogatives of the agencies concerned. These other uses of information become controversial in the context of privacy protection because they foster the collection of more personal information than is required for effective control of mass clienteles. Therefore, public surveillance represents but a single facet of organizational efforts to achieve control through the use of information originating in personal files.

Rule's reformulation of privacy issues as a theory of public surveillance represents an improvement on the privacy literature's characterization of contradictory public demand for a rational allocation of services which must not, however, impinge upon privacy. His theory is less adequate as a comprehensive theory of data-based social control associated with computer information systems because (1) it confines the phenomenon of social control to the agency/mass clientele relationship; (2) it does not acknowledge other contexts within which service institutions may use information for purposes of social control (planning, social forecasting, etc.); and (3) it does not acknowledge similar social control activities by nonservice firms. In short, his theory fails to account for a whole gamut of social-control activities historically associated with corporate use of the computer.

In this chapter, I have identified a fairly longstanding project to orient social research to the demands of the cybernetic economy. This project initially surfaced under a number of rubrics: social technology, social management, and social engineering. More recently (since the mid-1960s), it has been assimilated into futures research, which has occluded the negative connotations of social manipulation. Computer technology is vital to this project because it enables social researchers to simulate the activity of complex social systems, and, ideally, to predict their behavior. Computer data banks will function as enormous depositories for social data; computer-based home networking will be a direct conduit to these data banks. Advances in computer hardware, as well as in simulation technology, will improve both the accuracy and the value of these systems. It is important that the social significance of home networking be considered in the context of the cybernetic economy, and the necessity of social research for social management as a component of this economic order.

4.8 The Critique of Bureaucratic Rationality

In this chapter, social management and public surveillance have been examined at a practical level as components of contemporary organizational technology. My focus on home networking has directed attention to how these practices enable organizations to manage the environment external to the firm. At a higher level of social analysis, these practices must be considered in their relationship to the growth of bureaucracy and the extension of bureaucratic rationality as analyzed in classical sociology (Weber) and critical theory (Marcuse, Habermas). Weber's work, and to a lesser extent the exchange between Marcuse and Habermas on technical reason and ideology, are sufficiently "monumental" to the literature on this subject and do not need a full exposition. My brief review of this literature is limited to the critique of bureaucratic domination and to a consideration of what would be required, according to critical theory, if man were to free himself from this bondage.

For Weber, as for critical theory, the question of bureaucratic domination is framed by the problem of legitimation. To the extent that power is an intersubjective, social relationship in which both the ruler and the ruled accept the claim of the one to impose his will on the other, the problem for sociology is to determine how this power relationship should appear as legitimate to both ruler and ruled alike. Weber created a tripartite scheme for characterizing the "inner justifications" of "basic legitimations of domination" (1946, 79). These include traditional domination, charismatic domination, and legal domination. In modern society these legitimations of domination are intermixed, with the historical trend being toward legitimation on the basis of legal authority (i.e., consensus on the rightness of law). Bureaucracy is the typical apparatus for the administration of the legal type of domination.

Weber's ambivalent portrait of bureaucracy has provided fertile ground for controversy in scholarship. On the one hand, bureaucracy is the most efficient form of organization man has ever known. On the other, it is a major threat to individual freedom and democracy. To the extent that bureaucracy functions according to formalized rules administered by impartial functionaries, it has a leveling, egalitarian effect and may be considered essentially democratic. But the risk exists that the individual may become a victim of inflexible, too-rigidly-applied rules which threaten human freedom and spontaneity. Bureaucracy may be said to have abolished local and feudal privilege—but it has replaced this with the authority and privilege of the bureaucrat and the specialist. The modern bureaucratic apparatus is not easily subject to overt manipulation by the politician. It has become unresponsive to political authority. Most important, Weber established the superiority of bureaucracy as the type of administrative apparatus suitable to any system of political authority.

Although Weber may have been ambivalent about the repercussions of the rationalization process, he was firm about the superiority of bureaucratic rationalization and the inevitability of bureaucratic domination as the fated condition of human bondage. He notes:

> Precision, speed, unambiguity, knowledge of the rules, continuity, discretion, unity, strict subordination, reduction of friction and of material and personal costs—these are raised to the optimum point in the strictly bureaucratic administration, and especially in its monocratic form. As compared with all collegiate, honorific, and avocational forms of administration, trained bureaucracy is superior on all these points. And as far as complicated tasks are concerned, paid bureaucratic work is not only more precise but, in the last analysis, is often cheaper than even formally unremunerated honorific service. (1946, 214)

He continues:

> The ruled, for their part, cannot dispense with or replace the bureaucratic apparatus of authority once it exists. For this bureaucracy rests upon expert training, a functional specialization of work, and an attitude set for habitual and virtuoso-like mastery of single yet methodologically integrated functions. . . . More and more the material fate of the masses depends upon the steady and correct functioning of the increasingly bureaucratic organizations of private capitalism. *The idea of eliminating these organizations becomes more and more utopian* [emphasis added]. (229)

Marcuse's critique of Weber does not focus on bureaucracy per se, as an administrative apparatus, but on the underlying rationalization process of which bureaucracy is but the organizational manifestation. His critique of Weber's concept of formal rationality goes to the heart of the rationalization process and, by extension, exposes the particular bias of his thought. According to Marcuse,

Weber's technical reason of formal rationality was not formal at all. It was capitalist reason; Weber merely compounded the two. Weber's identification of formal rationality was in effect contingent upon "two very *material* historical facts," "the private enterprise" and "free labor" (Marcuse 1969, 222). To the extent that these acted as material preconditions of formal rationality, it is not formal. Formal rationality is historically determined by these forces of production, as well as by the unequal relations of production (i.e., domination) which underscore them. Marcuse states:

> But it is precisely here at this most decisive point, where Weber's analysis becomes self-criticism, that one can see how much this analysis has fallen prey to the identification of technical reason with bourgeois capitalist reason. This identification prevents him from seeing that not "pure," formal, technical reason but reason of domination erects the "shell of bondage," and that the consummation of technical reason can well become the instrument for the *liberation* of man. (223)

He adds:

> The very concept of technical reason is perhaps ideological. Not only the application of technology but technology itself is domination (of nature and men)—methodical, scientific, calculated, calculating control. Specific purposes and interests of domination are not foisted upon technology "subsequently" and from the outside; they enter the very construction of the technical apparatus. Technology is always a historical-social project: in it is projected what a society and its ruling interests intend to do with men and things. Such a "purpose" of domination is "substantive" and to this extent belongs to the very form of technical reason. (223–24)

From this critique Marcuse goes on to consider the utopian possibility of a new type of technical reason disengaged from the historical imperatives of capitalist domination. This new rationality would be qualitatively different: "Technical reason is the social reason ruling a given society and can be changed in its very structure. As technical reason, it can become the technique of liberation" (225).

Although Habermas agrees in principle with Marcuse's critique of technical reason as ideology, he presents a much more complex picture of the relationships among technical reason, domination, and legitimation in the modern state. As well, he rejects Marcuse's utopian formula for a new science and technology based on a partnership with nature. Habermas demonstrates that Marcuse's utopian claims posit a new attitude toward science and technology, without, however, positing a fundamentally new science or technology. Moreover, such a project is unrealizable. In order to understand why, it is necessary to review Habermas's distinction between work and interaction.

Habermas distinguishes between two systems of human action: work (systems of purposive-rational action) and interaction (communicative action). The former is concerned with technical rules abstracted from empirical experience; the latter involves socially established norms. In a further step, Habermas suggests that a society may be characterized by the interrelationship of these systems of action: for example, by which one predominates over the other. With references to Gehlen's work, Habermas argues that modern science and technology are immanently connected with human behavioral systems of purposive-rational action. According to Habermas, Marcuse is correct when he postulates a new relationship between technology and power in the modern state, but he is mistaken in his belief that one may liberate man by simply reorienting his attitude to science and technology. Instead, the problem is to understand a new configuration of systems of purposive-rational and communicative action, particularly as they affect legitimation. Whereas in earlier societies legitimation came "from above," from institutional frameworks of communicative action, more recently it has imposed itself from below as an extension of systems of purposive-rational action. Habermas reformulates Weber's concept of rationalization in the context of a tendency in modern societies for systems of purposive-rational action to displace communicative action as frameworks for a consideration of social norms and values. Subsystems of purposive-rational action colonize institutional frameworks with the promise that norms and values may be determined through procedures of rational action. An ideology of science and technology conceals this tendency by obscuring the fundamental difference between purposive-rational action and communicative interaction.

At one level of analysis, social management and public surveillance are expressions of bureaucratic administration as characterized by Weber. As such, they remain as problematic for society today as they were in the time of Weber. At another level of analysis, they are organizational manifestations of the rationalization process. Although Weber was the first to consider it in detail, this phenomenon has assumed new dimensions with the increased role played by science and technology in capitalist economic development. These developments are most perfectly understood in the work of Habermas, who identifies the particular dynamics of economic development, rationalization, and legitimation proper to the modern state as follows:

> What is singular about the "rationality" of science and technology is that it characterizes the growing potential of self-surpassing productive forces which continually threaten the institutional framework *and at the same time,* set the standard of legitimation for the production relations that restrict this potential. (1970, 89)

Clearly, an effective challenge to practices of social management must be mounted not at the level of a critique of bureaucracy (as a threat to personal autonomy and human spontaneity), but in terms of a critique of domination in the modern state as elaborated by Habermas, and, to a lesser degree, by Marcuse. As such, social management is a particular institutional expression of instrumental rationality understood as the displacement of systems of communicative interaction by systems of purposive-rational action. Although the consideration of home networking technology is several steps removed from this level of theorization, the critique of instrumental rationality is valid as a macrocritical framework for understanding these systems. Moreover, it reinforces the arguments presented in this chapter concerning the inadequacy of public surveillance as a framework for understanding the social-control implications of home networking.

Home networking will extend the administrative apparatus of corporate bureaucracy into the home more directly and more perfectly. This development coincides with the advance of bureaucratic rationalization and systems of purposive-rational action analyzed in Habermas' critique of instrumental rationality. Although understanding the social-control implications of home networking contributes little to the critique of legitimation per se, understanding the connection between interactive systems and the technical apparatus of bureaucratic rationalization is relevant to this critique.

5

The Political Implications of Home Networking and Social Management

5.1 Introduction

Home networking represents a single but vital facet of a more extensive transformation of the communications infrastructure associated with the information economy of late capitalism. The study of its political implications should therefore consider a number of trends in this economy. These trends characterize the current environment in several contexts (regulatory, communicational, political, and judicial) for the introduction of new electronic services for the home. Factors to be considered would include (1) an ongoing politics of deregulation regarding the telecommunications and broadcasting industries; (2) a progressive commodification of older communication forms as they become mediated through the introduction of new communications technologies, together with a generalized economic trend toward higher levels in the commodification of information; (3) a fragmentation and polarization of the political sphere along with the rise of interest-group and issue-oriented politics; and (4) the problem of guaranteeing rights of use in exclusivity (i.e., property rights) to information stored and transmitted in digital form.

5.2 Deregulation and the Telecommunications Industry

In recent years regulatory reform has emerged as a major objective of governments in the United States and Canada. Under a banner of deregulation, governments have reduced control over industries as diverse as banking, air transport, petroleum, and telecommunications. Although deregulation is a general trend, pressures for the reform of telecommunications, and the problems associated with its implementation, have been shaped largely by factors specific to the telecommunications industry itself.

Technological innovation is most often identified as the catalyst for change in the telecommunications sector. This change has, in turn, prompted the demand for a reform of regulations governing the industry. More specifically, the convergence of formerly distinct technologies, telecommunications and computing, has been identified as the principal source of regulatory uncertainty. The marriage of computers and communications has rendered current regulations inadequate and obliged policy makers to reevaluate the frameworks and categories that they have traditionally applied to the regulation of telecommunications.

In an article entitled "Computer or Communications?" Berman (1977) succinctly describes the dilemma of regulators faced with hybrid systems, combining elements from formerly distinct computer and communications technologies. Traditionally, governments have intervened to regulate telecommunication carriers where granting monopoly status to a single carrier is seen to be in the public interest. The argument is that a single carrier is better equipped to offer service at a reasonable rate to a majority of consumers than a plethora of competing companies with duplication of resources and costs. Although the primary business of telecommunications is the transfer, or carriage, of signals from point to point along a network, such an operation frequently involves a number of complementary functions, which, in the jargon of telecommunications regulation, are referred to as "incidental" to carriage. These would include, among others, "the encoding and decoding performed by the telephone instrument or computer terminal, the directory assistance service, and billing" (151). Because these services are essential to the overall functioning of telecommunications, they have traditionally been regulated alongside the essential carriage function of the monopoly firm.

The mandate of regulatory commissions, such as the Federal Communications Commission (FCC) in the United States and the Canadian Radio-television and Telecommunications Commission (CRTC) in Canada, restricts their jurisdiction to the broadcasting and telecommunications sector. They cannot intervene in the computer or data processing industry. But, as Berman points out, many of the so-called incidental services, which have been regulated in conjunction with monopoly offerings, are essentially data processing activities. The advance of computer technology has enabled monopoly carriers to automate a number of incidental functions. By the same token, however, cost reductions associated with computer automation have encouraged external firms to offer these same functions (or modified versions of them) as new services to compete with the telecommunications companies. Berman illustrates this point with references to three new services: "separately charged directory assistance, '911' or emergency calling with Automatic Location Identification, and packet switched computer networking" (152). To determine whether these services should be regulated or not, regulators have to distinguish between communications and data processing. It is impossible to do this on the basis of the technologies involved:

> Technological advances in communications and data processing have not "cooperated" with the law. As computer and communication services have developed, they have become dependent on similar technologies, bringing their different market structures into contact. The line separating these services—itself based on technological considerations—has become as much a barrier to understanding as an aid to decision. (150)

In the case of telecommunications, deregulation is taking place against a background of a regulatory crisis which has called into question the very legitimacy of traditional regulation itself. Because frameworks for regulation have been designed according to an earlier "transportation" model of communication proper to the regulation of land transport, they do not apply to new telecommunication services which typically must integrate data processing and communications as a condition for introducing them.

In his 1983 book entitled *Technologies of Freedom,* de Sola Pool has formulated one of the most systematic arguments in favor of deregulating telecommunications. His major premise is that computer technology is moving to supplant the printing press as the favored instrument for the expression of ideas. Accordingly, he asks: "If computers become the printing press of the twenty-first century, will judges and legislators recognize them for what they are?" (189). De Sola Pool contends that maintaining current telecommunication regulations in an era of electronic publishing and computer networking will amount to a form of censorship. His arguments against regulation are essentially based on constitutional, First Amendment considerations.

De Sola Pool identifies three communication structures, each regulated according to a different model: "the print model, free from regulation in general; the common carrier model, with government assuring nondiscriminatory access; and the broadcasting model, with government-licensed private owners as the publishers" (233). According to de Sola Pool, the model developed for broadcasting was fatally flawed and probably unconstitutional. It did nothing to foster freedom of expression, and probably impeded it. "To the detriment of freedom, however, neither the option of common carriage nor that of the market was chosen for radio and television policy. American radio and television has settled into a regime regulated by the values and judgments of public authorities" (149).

Regulation of the broadcasting industry has traditionally been justified on the basis of channel scarcity. De Sola Pool contends that the advent of new media such as coaxial cable, fiber optics, videodisks, and satellite dishes has effectively removed this obstacle to free expression. If the broadcasting model represented an unconstitutional impediment to free expression in an era of scarcity, it is clear, according to de Sola Pool, that its application to telecommunications in an era of channel abundance cannot be justified. De Sola Pool's recommendations, inspired in large measure by neoliberalism, favor a modified version of the "free market" print model. The neoliberal, laissez-faire approach is best exemplified in the ten principles which de Sola Pool argues must guide future telecommunications policy. Only his fifth recommendation is a weak accommodation to the traditional common carrier model. These principles represent a coherent synthesis of the deregulatory stance. In abridged form, they are as follows:

> (1) The First Amendment applies fully to all media. (2) . . . anyone may publish at will. The core of the First Amendment is that government may not prohibit anyone from publishing. There may be no licensing, no scrutiny of who may produce or sell publications or information of any kind. (3) . . . *enforcement of the law must be after the fact, not by prior restraint* [emphasis added]. (4) . . . *regulation is a last recourse* [emphasis added]. In a free society, the burden of proof is for *the least possible regulation of communication* [emphasis added]. (5) Interconnection among carriers *may* [emphasis added] be required. The basic principle of common carriage, namely that all must be served without discrimination, implies that carriers accept interconnection from each other. (6) . . . recipients of privilege may be subject to disclosure. (7) . . . privileges may have time limits. Patents and copyrights are for finite periods, and then the right expires. (8) . . . the government and common carriers should be blind to circuit use. (9) . . . bottlenecks should not be used to extend control. (10) . . . for electronic publishing, copyright enforcement must be adapted to the technology. (246–49)

At first glance, de Sola Pool's recommendations appear to represent an enlightened manifesto for communications in a computer age. More careful scrutiny suggests, however, that his analysis may be seriously flawed. De Sola Pool is correct in noting that computer technology will play a major role in publishing in future decades. But the "computer equals printing press" formula is too restrictive. The computer is also an accounting tool, a simulation tool, a surveillance tool, and an automation tool with applications in many industries at many levels of work. The traditional printing press can perform none of these functions. When the computer is combined with telecommunications, its impact on communications is not restricted to an automation of the publishing function. As noted in the previous chapter, the accounting function makes it possible to record who receives what information and when, with all the repercussions for privacy and social management that this implies. Digital technology opens the door to a commodification of communications on a level unanticipated by the liberal theory of a free press. These issues are examined in more detail later. For the moment, it is sufficient to view de Sola Pool's arguments as representative of the deregulatory stance in the United States.

For reasons specific to Canada's geography (its proximity to a much more populous United States) and its federation system of federal/provincial jurisdiction (which decentralizes considerable power in the hands of provincial governments), deregulation has been pursued less enthusiastically. A recent report prepared by the Institute for Research in Public Policy entitled *Regulatory Reform in Canada* (Stanbury and Thompson 1982, 22) laments, in particular, the lack of interest in deregulation expressed by Canadian business. As the title of the report suggests, in Canada the accent has been placed on reform, as opposed to deregulation. Technological innovation, mentioned earlier in the U.S. context, is also identified as the catalyst for reform in the telecommunications sector. But in Canada the goal of regulation, related to freedom of expression, has traditionally been supplemented with national unity and national sovereignty objectives. This means that any attempt to deregulate telecommunications in Canada will be tempered by national sovereignty considerations.

Jurisdictional overlapping, combined with the fragmented responsibility and nonstandard criteria associated with partially decentralized Canadian regulatory structure, poses considerable problems for companies seeking to introduce new services in Canada. (Schmidt and Corbin 1983, 219). This structure has also made it difficult to coordinate and plan the development of the telecommunications at a national level. In a recent article entitled "Telecommunications in Canada: The Regulatory Crisis," Schmidt and Corbin have assessed the Canadian system in the following critical terms:

> Canada is a balkanized telecommunication jurisdiction, in which the attempt to gain reasonable access to the Canadian telecommunications system as a whole is a costly, time-consuming and frustrating effort—and usually only partially successful. The regulatory configuration runs counter to the logic of Canadian federalism, and to the nation-building role that the country's major transportation and communications systems have historically played. (225)

In spite of these factors, the Canadian regulatory response to new services has followed a course parallel to that of the United States. In 1980, the CRTC allowed the attachment of customer-provided terminal equipment to the public telephone network. In 1982, Canadian National/Canadian Pacific Telecommunications (CNCP) introduced its Econovoice service, a private line voice system in competition with the telecommunications companies. More recently, CNCP has requested interconnection with the public switched telephone network. Although the request was initially refused by the CRTC, the dossier is currently under review by the Department of Communications in Ottawa. A similar trend is under way for a loosening of controls over the cable industry in Canada.

To summarize briefly, the introduction of new communication services to the home is closely related to changes in the North American structure for regulating telecommunications. Home networking is a part of this trend. Its implementation follows logically from precedents established in terminal attachment, system interconnection, the breakup of AT&T (which opens the way toward its entry into unregulated markets) and a general lessening of controls over the cable industry. These developments make it economically and politically feasible to propose new services which until recently have been only technologically feasible.

The restructuring of communication regulations is, in effect, a precondition for the introduction of home networking. But regulatory change is not occurring in a neutral political environment. That adjustments to the regulatory structure are being justified by a political ideology of deregulation is significant. Deregulation, when applied to telecommunications, implies more than a simple reform or readjustment of regulatory policy, as de Sola Pool's stance in *Technologies of Freedom* illustrates. He defends free speech and constitutional rights at a rhetorical level in order to justify a removal of controls from the communications industry at the very juncture where more vigilance is required.

The computer and telecommunications industries are both currently dominated by very large firms. As more and more communications take place through computer networks, these networks will play an important role in the circulation of ideas; directly affecting the formation of public opinion and social decision making. As Mowshowitz (1983) has noted, "the challenge to regulators comes from the *closeness of communication carriers and information providers* [emphasis added], which may be divisions of the same company or have close commercial ties, and the power of the dominant companies in the computer-communications sector" (7).

In the first chapter, I mentioned that only very large firms will be able to support the financial burden of implementing new services in the home. Difficulties in creating markets for new services will, therefore, foster conditions conducive to forms of monopolistic and oligopolistic industry control. De Sola Pool's deregulatory proposals, based on the faulty analogy of the computer as printing press, ignore this reality:

> Monopoly, however, will not be the usual situation in electronic publishing. If access to the basic carrier exists, entry is easy to the electronic information market. Electronic publishing over common carriers may be as competitive as print publishing or more so. (1983, 211)

In the light of current developments, it would be difficult to argue against regulatory reform in telecommunications. But de Sola Pool's stance is grounded in the ideological preoccupations of neoliberalism, instead of in an analysis of the technologies and industry structures involved. It runs counter to objectives of free expression. Deregulation is a political stance, inspired by a neoliberal ideology. It is not equivalent to reform. When applied to telecommunications, it does little to advance public understanding of the new communications infrastructure and of the challenge it presents to society. When deregulation is uncritically linked with goals of free speech, it confuses issues and ultimately undermines its stated objectives.

5.3 The Commodification of Communications

Home networking is linked, at a technological level, to the expansion of computer networks outward from the corporate sector. From an economic perspective, home networking is connected with an equally significant, if not more important, trend toward the commodification of communications. Although commodification is not new to communications, the advent of microelectronic technology has made it possible to transform communications activities into a gamut of new services, which otherwise could not have been exchanged in the marketplace. The various forms of electronic teleconferencing illustrate this point. Face-to-face encounters have given way to audio, video, and computer teleconferencing. Communications technology has fostered a progressive transformation of group work into an electronic activity with its own price tag.

This trend is significant for several reasons. From a macroeconomic perspective, the ability to create new products and generate new markets for them is a boon to the economy. This is the main significance of the so-called microelectronics revolution, for without this capability the conditions of perpetual expansion and economic growth so essential to the longevity of capitalism would cease to exist. The "miracle" of microelectronics lies less in the ability to miniaturize circuits than in the economic potential which this miniaturization creates for an almost limitless reinvention of commodities.

The repercussions of this trend extend beyond its benefits to the economy of late capitalism and the interests of capital. New electronic services do not represent a simple repackaging of older activities. They frequently imply a radical reorganization of communicational patterns; they may do this to such an extent that it may often be more appropriate to speak of entirely new systems of communication. A primary example of this would be computer teleconferencing (Kerr and Hiltz, 1982).

In addition to producing qualitative changes in communication, advances in microelectronics have fostered an almost exponential growth in the types and number of exchanges which can now be mediated through communication systems. In addition to transforming older communication services and creating new ones, communications technology will deliver and ultimately transform other services in such areas as retailing, security, and banking (i.e., teleshopping, telesecurity systems, telebanking). The commodification trend will, therefore, affect communications by (1) modifying existing communication services and creating new ones, as well as by (2) introducing a "communications" component to services which are not presently communicational at their base.

Greater commodification of communications has an obvious significance for the electronics and communications industries. But what will be the significance of this trend for the whole of society? Such a question is intrinsically linked to society's perception of the function and value of communications. Historically, the political sphere has been the privileged context for a public debate on communications. The origins of this debate may be traced to the rise of liberalism and the struggle for freedom of expression which accompanied it in the Western democracies. Two important outcomes of this struggle were the rise of the press as an established institution of democratic societies and the emergence, at a philosophical level, of the liberal theory of the press grounded in liberal political ideology. Although liberalism has undergone significant modifications and revisions as it confronts the reality of mass society, the philosophical justification for the freedom of the press has remained relatively intact. Moreover, this same theory, designed as a justification for an emerging institution of the press, has underscored debate and policy regarding communications in other contexts.

In his *Liberalism, Democracy, and the Mass Media,* Jensen (1976) has aptly characterized the liberal theory of the press as "a species of social myth grounded in the *Weltanschauung* of Classical Liberalism" (182). By this, he implies that there can be no definitive proof of the theory. If a consensus exists about the theory, it is because of tradition, political expediency, and institutional (i.e., the press) self-interest. In the following passage, Jensen elaborates on the principal tenets of classical liberalism as they apply to a theory of a free press:

> The Liberal doctrine of press freedom rests also upon a strong faith in the *self-righting process,* upon a faith that *truth will emerge from the clash of knowledge and opinion in a free and open market* [emphasis added]. It assumes that man is primarily a rational being; moreover, *that man is disposed to seek truth and to be guided by it* [emphasis added]. Using his reason, man can find truth, or at least a considerable and useful portion of it, so long as he has access to all knowledge and opinion. The doctrine also recognizes that no one person or group is wise enough to be trusted to discriminate between valid and invalid ideas, that suppression of itself tends to impart a fictitious validity to the suppressed idea and that, in any event, an idea if valid will in the long run survive such suppression. Even traffickers in lies and distortion must be free to have their say. If speech or publication is mere fulmination, it will be ultimately impotent. As men find truth in the free play of ideas, social change will result from the non-disruptive process of conversion instead of from revolution. (62)

Although focused on a particular struggle of the times (i.e., censorship by the Crown of an emerging press), this blend of epistemology, politics, and economic self-interest has proven extremely durable. Notwithstanding its questionable epistemological assumptions, the doctrine has been extremely influential at a political level, where it continues to serve as a consensual basis for policy formation in an era of electronic communications. The notion that all ideas must be permitted to circulate in a free marketplace, where they will be exchanged and ultimately "purchased" or rejected according to their relative value, is intrinsic to our political system (Becker 1945, 29–30). Moreover, this market model of the political exchange of ideas has become, by default, our model of communications in the public sphere.

What, according to liberal theory, are the implications of the commodification trend in communications for the political sphere? First, it is important to remember that the commodification of political communications does not originate in computer-mediated communications. Political campaigns, particularly at the national and state/provincial levels, frequently employ a mix of electronic (primarily broadcasting) and print media to create a candidate's "image" and communicate this to the public. Access to such media is a matter of economics and financing. Although it is impossible to buy an election with media exposure, a successful campaign has financial resources at its disposal which would be adequate to the task of competition in a media-"mediated" campaign. The free competition of ideas in such a context is clearly not "free," because the use of media implies that political exchange has a cost. Of course, political communication is not solely a function of political campaigns and government management of the media agenda. Media bias, both structural and personal, may also represent a constraint on the free exchange of ideas. Particulary troublesome for liberal theory is the case where editorial politics are determined by the politics of ownership.

In order to rectify this contradiction, or at least minimize its potential repercussions for liberal theory, the Western democracies have opted to impose limited, piecemeal controls on those industries involved in transmitting and disseminating information deemed essential to the public sphere. Chiefly, governments have intervened to discourage monopoly ownership in the telecommunications industry. Where monopoly has appeared to be "natural" (i.e., in the public interest), regulatory oversight (FCC, CRTC) has been instituted. In the case of broadcasting, instruments such as the fairness doctrine have been introduced in order to ensure equal access.

The primary thrust of this intervention has not been to positively reinforce a free exchange of ideas as much as it has been to negatively discourage the generalization of conditions under which the liberal pretense of free exchange might appear as an empty facade. There is, for example, no compelling reason to believe that oligopolistic control of an industry by two or three very large firms will be any more conducive to ideological diversity than monopoly ownership by a single corporation. Similar class interests will be represented and protected, regardless of the favored model.

Such is the liberal theory of the press as it has attempted to maintain its legitimacy in the face of numerous challenges and contradictions posed by an economy of late capitalism. To the extent that liberal theory has always incorporated elements of the economic model ("exchange," free "marketplace," and so on), commodification does not pose any special problem to the theory. Free exchange does not imply that ideas—or, more specifically, their circulation—should be free in any economic sense of the term, but rather that the enunciators of these ideas should be at equal positions of strength as they approach the marketplace of expression.

The potential of computer-based communications to break down communication services into highly discrete units, and to monitor automatically their distribution to consumers, will engender levels in the commodification of information heretofore unimaginable in older forms of public communications and unanticipated by the liberal model. The case of a traditional newspaper versus its electronic counterpart, accessed through a computer terminal, illustrates this point. When a consumer buys a newspaper, he buys a total package of heterogeneous information. From the consumer's point of view, commodification is a factor of exchange: 35 or 50 cents for a total package of information and entertainment. When this same information is accessed in an electronic database provided by the same newspaper, a radically different dynamics of commodification and exchange is engendered. First, communication fees are levied by the telecommunications carrier. This is done along two axes: according to the number of bits transmitted through its lines, and according to the period of time the user is on the system. Access to the database itself requires payment of an initial subscription fee which varies (in most cases) between \$50 and \$150. Once connected to the database, the consumer is charged according to the quantity of information viewed (measured in terms of pages, words, or bits) and/or according to the time he spends on the system.

An emerging model of exchange and interaction with firms producing, packaging, and distributing information is based on the potential to itemize costs to the so-called end-user. Digital forms of communication make it possible to subdivide the numerous aspects of new services into discrete functions, to itemize their costs, and to ultimately pass them on to the consumer.

Although a transition to this new model is just commencing, there are indications of some resistance to it. The failure of the initial videotex projects is a case in point. The electronic database described above is far from being an electronic newspaper. Conspicuously absent from this type of experience is the entertainment function so essential to the packaging of information in a newspaper format. The early videotex projects seriously overestimated public need for information in its pure form, stripped of the entertainment function. The use of graphics, colorful images composing themselves in layers upon a video screen, was the only apparent accommodation to entertainment foreseen by the project. The consumer balked when confronted with a model of information and news consumption which attempted to transform a type of ritual behavior, with an implicit entertainment function into an information "service" with itemized costs. There was little entertainment value in the proposed services, the information was not portable, and, most important, it was available at a lower cost from traditional sources.

These problems are in part a function of the technology (i.e., related to the limits and particular traits of the technology employed) and in part the result of inadequate marketing. A more fundamental obstacle to the new services is that the average citizen does not appreciate the instrumental value of information. He does not perceive information as a means toward the realization of personal goals. When a consumer consults the mid-week edition of a newspaper for sale items, this activity is relatively transparent to him. It is not viewed as an information-gathering activity, let alone as a "service." He probably does not gauge the cost of this information relative its economic returns. To a large extent, the consumer's ignorance of what he is doing—and, by implication, his ignorance of the value of information to this activity—is a function of current practices in the packaging of information. As long as information of various kinds (political, real estate, stock market, weather, entertainment, and so on) is bundled together in an undifferentiated form, and financed primarily through indirect subsidies derived from advertising, it will be difficult to market the new information services. Their success depends upon the indoctrination of the individual to a new consumer attitude that information has an intrinsic value. Only when this has been achieved will it be possible to introduce, on a large scale, a class of specifically informational services whose commodification will be accepted equally with that of material goods and noninformational services.

Such a transformation of consumer attitudes will not occur overnight. It will happen in perhaps five to fifteen years from now. But the process has begun on a scale sufficient to merit recognition as a major trend with significant repercussions for the future of communications. Current examples of this development would include the introduction in the United States, in the wake of the AT&T divestiture, of "measured" telephone service; pay TV; initiatives by the cable industry to develop "pay-per-view" programming; the scrambling of specialty programming distributed by satellite (such as HBO) to discourage "free" capture of signals by Television Receive Only dishes (TVROs); attempts by the satellite scramblers to market authorized descrambling devices; the gradual but steady emergence of commercial networking services such as The Source in the United States; and the deregulation of cable TV and the concurrent relaxing of controls on cable rates.

Just as significant are indications that the process of educating the public to this new reality has begun in earnest. For the moment, "information equals power/ productivity/ profits" is the formula integral to efforts to market computers and computer communications in the business sector. This is the thrust of advertising directed at the business consumer in professional and commercial journals, as well as in the mass media (e.g., the Chaplin ads by IBM). The formula is credible in a business context, with the important qualification that business is increasingly aware that gains are not automatic. It is essential to carefully manage the integration of information systems into organizations.

In order to market the new services to a nonbusiness, mass clientele, it will be necessary to modify the "information equals power, productivity, and profits" formula to one along the lines of "information equals leisure, financial security, savings, status, opportunity, and personal fulfillment." This will represent a considerable challenge to marketers. I cannot empirically evaluate the extent of current efforts in the mass media to create this new consumer mentality, but it is reasonable to anticipate, given the stakes involved, their eventual success. A great deal of terrain has already been prepared in the context of the widely held belief that we are living in an information society. According to this concept, information and knowledge are not only the basis for a restructuring of the economy, but are also factors of social progress and social evolution. It requires no great leap of faith or cognition to derive from such an assumption the belief that knowledge and information may lead to personal advancement and fulfillment.

The political repercussions of commodification and home networking should be reevaluated in a more comprehensive critical context. In a recent article entitled "The Media and the Public Sphere," Garnham (1986) has succinctly identified the contradiction inherent in the liberal theory of a free press. He states:

> The problem with liberal free press theory is not just that the market has produced conditions of oligopoly which undercut the liberal ideal nor yet that private ownership leads to direct manipulation of political communication, although it does, but that there is a fundamental contradiction between *the economic and the political at the level of their value systems and the social relations which these value systems require and support* [emphasis added]. (31)

He adds that

> in the political realm the individual is defined as a citizen exercising public rights of debate, voting etc. within a communally agreed structure of rules and towards communally defined ends. The value system is essentially social and *the legitimate end of social action is public good* [empha-

> sis added]. But within the economic realm the individual is defined as a producer and consumer exercising private rights through purchasing power in pursuit of *private interests* [emphasis added], his or her actions being coordinated by the invisible hand of the market. (31)

Garnham argues, with references to Habermas' ideal speech situation, that it is essential to critically "re-evaluate the public service model of public communications" on the basis of Habermas' work on the public sphere. According to Habermas, the creation of a public sphere was fundamentally related to the interests of a new political class, the bourgeoisie. The infrastructure of the newspapers, publishing houses, libraries and universities established in the eighteenth century emerged as a response to the collective political interests of this class. The public sphere was, on the one hand, protected from the power of the church and state, and, on the other, independent of pressures exerted by civil society. Historically, "the space between civil society and the state which had been opened up by the creation of the public sphere was squeezed shut between these two increasingly collaborative behemoths" (Garnham 1986, 30).

Combining Habermas' arguments concerning the fall of the public sphere and Garnham's ideas about the inherent contradiction between political and economic value systems, it becomes apparent that the political implications of home networking are best understood in the context of a crisis of the public sphere. It is clearly inappropriate to limit a discussion of the political to a consideration of political campaigns, political polling, electronic referenda, and their impacts, as has been the trend in the literature on home networking (e.g., Pergler 1980; Parkhill 1979; Tydeman et al. 1982).

Higher levels in the commodification of information, made possible by home networking, represent a major colonization of the public sphere by economic forces. To date, the public debate on mass communications and the public sphere has focused primarily on broadcasting technology. The challenge has been to discourage the further intrusion of economics into the realm of politics. The state of technology has itself been a constraint upon efforts to commodify information and communications in the political sphere. In the case of broadcasting, technology was not computer-based. Access to information was not, therefore, susceptible to the forms of hypercommodification identified earlier. Telecommunications, which has been the domain of public service monopolies and regulated private monopolies, has traditionally been offered at rates accessible to all but the poorest citizens. But in the slowly emerging era of computerized communications there will be no "free lunch." The ability to be informed on political matters as well as the ability to communicate and to organize will be a function of access to the emerging network marketplace.

Computer technology threatens the uneasy equilibrium between the economic and the political which has been the hallmark of liberal free press theory.

It does this by enabling companies to itemize information and communications, to break them down into highly discrete units, and to monitor their consumption cost-effectively using automation technology. It is possible to know who received what information and when, as well as to know who communicates (i.e., uses the system) for how long and for what: telebanking, teleshopping, teleconferencing, and so on.

Even more important, however, commodification implies that economic values, grounded in priorities of individual self-interest and goals of capital accumulation and profit (derived from unequal positions of strength in exchange), will come to dominate the sphere of social intercourse. In a society already characterized by the hegemony of economics in social discourse, it becomes increasingly difficult to protest against this. On what basis—anthropological, psychological, philosophical, or other—may one object? Liberal theory, with its affinity for economic models and rationales, is easily coopted to further justify commodification.

The work of Williams (1961) is a counterweight to the economic model of communications. Williams, a specialist in the cultural studies approach to communications, underscores the important point that human communication cannot be understood as a simple transfer or exchange of messages between two points. In his view, communication is a process whereby we give meaning to our experience, and through the sharing of experience participate in the formation of community. He states:

> Communication is the process of *making unique experience into common experience* [emphasis added], and, it is, above all, *the claim to live* [emphasis added]. For what we basically say, in any kind of communication, is: "I am living in this way because this is my experience." The ability to live in a particular way depends, ultimately, on acceptance of this experience by others, in successful communication. Thus our descriptions of our experience come to compose a network of relationships, and all our communication systems, including the arts, are literally parts of our *social organisation* [emphasis added]. (55)

Williams points out that

> the selection and interpretation involved in our descriptions embody our attitudes, needs and interests, which we seek to validate by making them clear to others. . . . Since our way of seeing things is literally our way of living, *the process of communication is in fact the process of community* [emphasis added]: the sharing of common meanings, and thence common activites and purposes; the offering, reception and comparison of new meanings, leading to the tensions and achievements of growth and change. (55)

If, as Williams argues, communication is intrinisically related to the sharing of experience, the formation of communities, and our "claim to live," is it appropriate to characterize it in economic terms? What will commmunity look like in an era of the hypercommodification of communications? Are the goals of community associated with the sharing of experience consistent with the economic rationale of an electronic marketplace? These questions need to be asked in connection with home networking. They do not emerge easily from the discourse of liberalism—in which answers will be even more difficult to find.

5.4 Communication and the Crisis of Pluralism

The political sphere is often one in which competing social interests are mediated. According to classical liberal theory, the interests of individual citizens are negotiated in the political system. As an accommodation to the reality of mass society in an industrial era, liberal theory has acknowledged that the unit of political interest and intervention is no longer the individual. *Pluralism* is the term most frequently employed to describe a shift to the interest group as the unit of political action. This revision enables liberal theory to contend that democracy still functions, albeit through a higher-level interaction of groups of individual citizens.

For postindustrialists such as Bell (1973), pluralism is a double-edged sword. Postindustrial society is characterized by a pluralist political system which may be portrayed as democratic. But Bell is disturbed by the path that pluralism has taken. According to him, democratic participation has gotten out of hand.

> In fact, there is probably more participation today, at the city level, than at any other time in American history. But the very increase in participation leads to a paradox: the greater the number of groups, each seeking diverse or competing ends, the more likelihood that these groups will veto one another's interests, with the consequent sense of frustration and powerlessness as such stalemates incur. . . . Without consensus there is only conflict, and persistent conflict simply tears a society apart, leaving the way open to repression by one sizeable force or another. (160)

Bell contrasts the technocratic reason of the technostructure with the excesses of the political sphere. In the complex social structure of postindustrialism, it is essential, according to Bell, that the technostructure play a leading role in "guiding" society. In Bell's postindustrial scheme, the technostructure is associated with the forces of rationality. The political sphere, although necessary, is problematic because it is fundamentally irrational. Bell argues that

> the central point about the last third of the twentieth century . . . is that it will require more societal guidance, more expertise. To some extent, this is an old technocratic dream. But an earlier technocratic visionary like

> Saint-Simon felt that in such a technocratic society politics would disappear since all problems would be decided by the expert. (263)

If there is any doubt concerning Bell's position, it is clarified in the following statement: "In the end, however, the technocratic mind-view necessarily falls before politics. The hopes of rationality—or, one should say, of a particular kind of rationality—necessarily fade. . . . Politics, in the sense that we understand it, is always prior to the rational, and often the upsetting of the rational" (365).

In order to understand the political context of communications in future decades, one must take into account the opposition, identified by Bell, between political pluralism and technocratic guidance. Communications technology and new communication services will be at the center of this debate.

In the previous chapter, the technocratic position was identified with social management. No doubt, proponents of technocratic reason will present social management as an alternative to the excesses of pluralism. The debate on pluralism will, however, focus on issues other than that of technocratic reason. The defense of pluralism is closer to the interests and preoccupations of liberalism (i.e., maintaining the liberal myth of democracy). The natural enemy of pluralism is not technocratic reason but *corporatism* (Schmitter 1979), or, to cite another example, *Friendly Fascism* (Gross 1980). Proposals suggesting corporatism or even fascism as more appropriate characterizations of the social and political order represent a greater threat to the pluralist position.

Given these observations, it is relevant to demand whether the opposition identified by Bell represents, in fact, an appropriate interpretation of the social forces at play. If not, what is the social reality beneath Bell's analysis? What, if anything, is its significance for an understanding of the political climate in which home networking will be introduced? If technocrats plead too much pluralism while pluralists fend off corporatism and fascism, what is the relationship between pluralism and technocratic reason? Are they, in fact, opposed?

Beneath the surface of pluralism is the reality, not simply of interest-group politics, but of interest politics initiated and managed by large corporate units. If mediation is occurring in the political sphere, it is related mainly to the resolution of conflict between competing corporate interests and between these interests and the state. Domhoff's model of the policy process sheds considerable light on pluralism at the corporate level. In his view, the policy process has five steps: "*Resources* fund *research* that shapes *decision making*; the latter wins popular approval through *opinion making* and is enacted by *law making*" (Mosco 1982, 24). Resources, controlled for the most part by very large corporations, are the motor which drives the process. Media of communication are essential to the process of forming public opinion once decisions have already been made.

New media of communication will have a twofold impact on corporate-level pluralism. On the one hand, more channels and fragmented audiences will make it increasingly difficult to reach the public. Choice in mass media implies that consumption becomes more personalized, and media campaigns are less likely to reach their intended targets. In concert with this trend, the merging of microelectronic technology and telecommunications makes it possible for grassroots organizations to mobilize. Automation technology, such as desktop publishing, lowers the cost of organizing for political intervention. This would be the optimistic scenario as portrayed by de Sola Pool. On the other hand, the corporate sector's ability to monopolize public attention through the control of limited channels is diminished. This loss of control is magnified by greater citizen access to new media.

But this is not the whole picture. According to Domhoff's scheme, it is primarily the fourth and penultimate stage of corporate pluralism that will be affected by the new media. The first three dimensions are, however, the decisive ones for control of the political process. The multiplication of channels may render opinion formation more costly and chaotic, but it will not bear directly on the determinative research- and decision-making stages. The monitoring capabilities of the new systems will probably improve social research, and ultimately decision making, by improving the quality of social data. The pluralism debate therefore emerges as something of a red herring.

In the context of deregulation, pluralism functions as both a goal and a justification. Removing government control over the communications industry is designed to foster greater pluralism. That democracy still functions under the pluralist model encourages continued expansion of communications systems, which have largely shaped the contemporary form of pluralism. When the charge of too much pluralism is raised in the context of new technology, it no doubt refers to problems of opinion formation in an era of manifold communication channels. It is difficult to evaluate the extent of such a threat to corporate power. But pluralism would represent an intermediate stage in a longer process of technological innovation which has favored a diversification of communication products and a multiplication of channels. When the so-called network revolution occurs, bidirectional communication will compound communication channels, generating a more manageable communications environment.

Although long-term perspectives are important, it would be foolish to ignore the current dilemma which the new communications revolution poses to corporate interests. It is precisely within this environment that networking services will be introduced. The fragmentation and polarization of the political sphere associated with the rise of interest-group politics will be worsened by the first wave of new communication services. Depending on one's perspective, this will lead to a more conflictual society—or a more open one. It is difficult to evaluate either claim, and it is particularly difficult to gauge the extent to which new communications services may affect these trends.

Both claims originate from the same source. The same interests which proclaim a more open pluralist society through new technology (i.e., de Sola Pool) lament the trend toward a more conflictual society (i.e., Bell), albeit through different spokesmen. What is the significance of this? At the very least, it implies that the debate on the political implications of new communications technology has been and will continue to be mired in confusion. In this respect, the discourse on the political implications of these systems is ingenious: it is possible to argue for deregulation as an instrument of greater pluralism and at the same time urge society to use the new technology to curb the excesses of pluralism. The rhetoric of pluralism merely clouds the issue. If pluralism is equivalent to a more fragmented political sphere (and not, as de Sola Pool proposes, to a revitalized democracy), it does not bode well for attempts to ensure a more democratic communications system in which access and privacy will be assured. As long as public interests are defined according to narrow sectarian concerns, it is doubtful whether a coherent public policy can emerge which will be commensurate with the magnitude of changes and challenges posed by the emerging communications infrastructure. Piecemeal solutions will not be sufficient. Global, consensual perspectives are required, and it is doubtful whether these can surface in a political system of corporate pluralism.

5.5 Property Rights

The computerization of communications implies new modes in the production of information. Opportunities for commodification are one dimension of this change. Commodification, however, is contingent upon assuring property rights to the new information services. Without such guarantees to producers of information, there is little incentive for them to offer the service.

Information which has been stored and transmitted in electronic form poses a special problem for traditional copyright law. Because electronic information can be transmitted and subsequently reproduced at low cost, unauthorized third-party firms can resell this information. Computerized information is ephemeral, making it difficult to trace these activities. It is also susceptible to manipulation and restructuring, and may be repackaged and sold as a new service or product at a very low cost. When this occurs, it may be difficult to establish the copyright obligations of the repackager to the initial producer.

In an era of electronic, computer-based communications, the physical transmission of information represents a small fraction of production costs. Processing and packaging information for final presentation represent a further portion of overall costs, but the major investment for an information provider (such as a wire service) is gathering information. New technology does little to reduce the costs of collecting information, but it does offer opportunities for competition in transmitting, processing, and packaging it. If, however, the property rights of the firm which has absorbed the initial cost of gathering information are not respected, secondary use becomes a moot question. New forms of compensation for authors, whether corporate or individual, will have to be devised. Inability to guarantee such compensation will delay the advance of new information services.

The question of property rights to computerized information is significant for home networking because a network marketplace cannot flourish without mechanisms for compensation. Ultimately, this problem will impede the implementation of home networking. Moreover, once copyright is extended to the new services, it will demarcate this information, raising the issue of private control over public information. This poses no major problem if private control operates in an environment in which numerous firms compete. Current trends, noted in the first chapter, suggest that this will not be the case. Only a few very large corporations are presently offering electronic information services at a profit—Reuters, Dun & Bradstreet, Dow Jones, and McGraw-Hill—and there are few indications that a future electronic information industry will be more diversified (Field and Harris 1986). The extension of copyright protection to new forms of electronic information is, therefore, a precondition for the network marketplace. More important, however, copyright raises the issue of private control and public information in the new electronic services industry for the home.

5.6 Social Management and the New Communications Infrastructure

The trends and issues discussed in this chapter have an obvious significance for an understanding of the new communications infrastructure. Deregulation is not equivalent to reform. It is a political agenda grounded in a neoliberal, laissez-faire approach to industry/state relations. The growth of the information sector of the economy is contingent upon both a restructuring of the telecommunications, information, and computer industries and the establishing of a new relationship between these industries and the state. Deregulation is the framework for these changes. Although it fosters conditions favorable to the growth of these sectors, it does little to assure that this growth will favor anything but economic objectives. Commodification is the manifestation of these objectives in social interaction. The commodification of communications and information implies the predominance of economic modes of action in social exchange. Copyright, the legal framework for commodification, seeks to guarantee property rights (and, therefore, profitability) in new systems of communication. What is the significance of this for society?

At stake is the integrity of public and private spheres of action. Information and communications are crucial factors in the health of both spheres. But new systems of information and communication do not respect these boundaries.

The liberal project to create a purely political sphere as a watered-down substitute for the public and to isolate it from economic forces is in retreat. The major issue arising from these developments is the private control of public information and the impact of this on what remains of our democratic institutions.

At first glance, social management appears to be at the margin of this issue. This is because new communication technologies have been primarily viewed as future media of mass communications. A particular version of the access question associated with the broadcasting model has dominated social reflection on new interactive services. According to Domhoff's model, the debate on access is appropriate to one of the final stages in the policy process, the formation of public opinion. When private control of public information is treated in this context, it is assumed that this is the crucial stage in the political process. But Domhoff's model suggests that the pivotal moment may occur much earlier, when resources are allocated to specific research activities. Moreover, it is precisely at this stage that information derived from interactive systems will be applied to research for social management. If social management does not appear on the agenda of public debate on new communications technologies, it is because these activities have traditionally not been subject to scrutiny in debates on mass media broadcasting. This is understandable because traditional mass media do not generate issues related to institutional research. As long as these technologies are perceived as extensions of broadcasting media, this fundamental misconception will persist.

Social management raises the question of access to information and communications, but it does so in a context removed from the traditional preoccupations of the debate on mass media. Broadcasting media pose the question of access at two levels: first, citizen access to media products, and second, public access to the media instrument itself as a vehicle for the expression of ideas. The advent of interactive media introduces a third dimension, access to information from the bidirectional channel. Because this information will result from general public contact with these systems, the public should logically be granted a measure of control and access to this information, but the trends identified in this chapter suggest that this will not occur.

Information generated by interactive systems will circulate in an environment characterized by privatization. In a political climate of deregulation, it is unlikely that governments will intervene to assure the public character of this information. A more likely scenario is one in which government institutions compete both among themselves and with the private sector for control of the informational commodity. The general trend toward commodification of information and communications will affect home networking. The duplicity of political pluralism which calls for a multiplication of communication channels and services in the interests of democracy, while calling upon these same systems to curb their excesses, obscures the fundamental issue of social management. Private control of this information is not an issue for the proponents of pluralism, for whom the health of the communication and information industries is ample proof of the health of democracy.

What can be done? It is clear that the application of interactive technology to the home will have significant implications for our society and particularly for our political system. Creating policy for these new systems in a context of deregulation, unchallenged commodification, and self-serving corporate pluralism will not generate enlightened results.

Two major options are available for policy makers seeking to manage the informational commodity. Although diametrically opposed, these options warrant consideration. The first option, inspired largely by the experience of privacy regulation, suggests that controls are needed in order to limit the collection of data and to clearly demarcate the routes it may follow—how it will circulate, how it may be used and by whom, and so on. This would apply to all information generated as a result of an individual's contact with these systems, whether it be retained in personal or statistical form. In order to be effective, such a proposal would need an administrative body with responsibility for monitoring compliance with the law. This approach requires that the private character of personal information be established at the point of initial contact with these systems. Thereafter, use of this information would be contingent upon the individual's consent. Such an approach would apply property rights in such a rigid way that information could not be appropriated freely by others, regardless of its form. Applied strictly, such controls would not only restrict corporate, commercial use of this information, but would also create significant barriers to its ever achieving the status of public information. Although extreme, this approach has the advantage of being consistent. It recognizes that information has the characteristics of a property, regardless of whether the proprietor is an individual or a corporation. This option need not impede commodification and may ultimately intensify it, attributing property rights to information when it is collected.

As menacing as this option may be for commerce, a still more radical alternative would be to go against the grain of commodification and declare all information public from the moment of collection. This would not prevent industry from packaging it for commercial use. It would simply mean that all social actors begin on an equal footing with equal access to this social data. In return for access to data collected by other institutions and firms, the data collector would have to open his database to others.

Both options involve costs. In the case of the first option, these would accrue from the expense of licensing thousands of databases and monitoring their activities. In addition, there would be the cost of administering compensation to all those individuals prepared to relinquish their privacy in exchange for benefits. The second option resembles a decentralized version of the information utility. Because databases would be decentralized in the hands of data collectors, compatibility among systems would be essential. Technical protocols for information transfer and standards for data storage would have to be established. Systems would have to be simple. Undue complexity would render them inaccessible to all but the very largest institutions. Given current difficulties in creating computer networks for the office, which would enable microcomputers to "talk" to one another and communicate with mainframes, the second option appears to be technically ambitious. The recent formation of the Corporation for Open Systems (COS) by such competitors as IBM, APPLE, and GTE Telenet with the mandate to create a "standard set of software rules that would let different brands of computers communicate" suggests that there may be room for optimism (Harris 1986, 110).

Although the first approach addresses the issue of personal data privacy, it does not resolve the problem of social management. The second option confronts the issue of social management but does little for data privacy in these systems. A comprehensive approach to social control and home networking must incorporate both options. It is imperative that future policy for these systems guarantee the data privacy of individuals while at the same time assuring the public character of statistical social data. Unfortunately, if these options are costly when taken individually, they will be even more so when applied together.

Fortunately, in a technological society there is always an abundance of technical solutions. In a 1985 article published in *Communications of the ACM* (Association for Computing Machinery), David Chaum, of the Center for Mathematics and Computer Science in the Netherlands, has proposed a cryptographic system for electronic transactional systems which would meet the privacy needs of individuals while respecting the security and control requirements of firms. In "Security without Identification: Transaction Systems to Make Big Brother Obsolete," Chaum outlines a proposal in which the individual is given a "digital pseudonym" for each organization he deals with. These pseudonyms cannot be linked, but they do meet the control requirements of the individual firm or institution. The mechanism for conducting transactions is not centralized in the hands of the service provider, as is currently the case, but in a microprocessor which the individual holds in the form of a "card computer." Chaum's proposal is notable for the manner in which it respects the interests of both the client and the service institution. He states:

> The nature of the security provided under the new approach also differs substantially: current systems emphasize the one-sided security of organizations attempting to protect themselves from individuals; the new approach allows all parties to protect their own interests. The new approach relies on individuals keeping secret keys from organizations and

> organizations devising other secret keys that are kept from individuals. During transactions, parties use these keys to provide each other with specifically coded confirmation of the transaction details, which can be used as evidence of improper actions sufficient to resolve disputes. (1032)

Applying advanced cryptography to privacy in home networking merits consideration. If successful, it could lower the costs of state-administered privacy instruments. Along with such protection for the individual, it is crucial to insist upon the public character of statistical data generated by transactional systems. This information is a valuable social resource. It should be recognized as such by our society and incorporated into future practices for using the abundance of information and knowledge which is the promise of the new infrastructure of computer-based communications.

Conclusion

As an alternative to studies assessing the social impact of home networking, the critical thrust of my study has been to analyze its social significance within several distinct but interrelated contexts.

The first chapter has considered interactive systems in the microeconomic context of an emerging videotex industry. This analysis exposes the importance of the informational commodity as a vital component of a successful videotex industry. It also signals an important problem of home networking and society concerning new potentials for the expansion of systems of data-based social control. The literature on the social impacts of interactive systems suggests that this problem is at best poorly understood and documented, at worst totally overlooked. Economic and market analysis of the videotex industry reveals two potential social-control applications of the informational commodity. Only one of these, privacy protection, is an acknowledged concern of policy discourse on home networking. The other, which I have termed social management, is less well understood. In fact, it is not treated as problematic at all—at least not in the literature on home networking. It is, however, subject to controversy in critiques of systems analysis and futures research.

From my perspective, both theorizations of data-based social control are valid and highly relevant to an analysis of home networking. The consideration of home networking in the context of social management activities gained importance as it became apparent that the public was largely ignorant of the social-control implications of these practices. Particular attention has been devoted in chapter 2 to the legal-regulatory context of information control as the only major policy framework with a potential to intervene in and constrain uses of the informational commodity. The analysis of privacy statutes in the United States and Canada suggests a North American tradition of privacy policy which resembles a double-edged sword. On the one hand, statutes are significant because they represent an acknowledgement in law of the individual's legal, defensible right to control information about him in bureaucratic files; on the other, the cumulative effect of these statutes is not to substantially curtail bureaucratic prerogatives, but, instead, to legitimate them.

Chapter 3 has dealt with home networking in the context of actual and potential privacy policy. Privacy policy will not be an effective lever against corporate intrusions into the private affairs of citizens. This conclusion derives from a consideration of the North American tradition of privacy policy and future trends toward a transvaluation of the privacy interest into an economic interest subject to sale and exchange. Considerable weight and clarity is lent to this conclusion by Rule's work on systems of public surveillance, which effectively shifts the focus of analysis from a consideration of personal information abuse to a critique of bureaucratic surveillance. In modern social systems, personal documentation is the nexus of control for the bureaucratic management of mass clienteles. An effective challenge to these systems must go beyond a consideration of institutional abuse to confront the very legitimacy and social value of systems of public surveillance. This insight suggests that home networking is not a marginal problem which may be dealt with through ad hoc measures. A strong defense of privacy rights in home networking will be contingent upon efforts to bring the larger issue of systems of public surveillance to forums of public debate, to inform the public that abuse is only one objectionable facet of these systems, and to challenge the extension of these systems as an unjustifiable intrusion of the corporate sector into the private lives of citizens.

But the application of the informational commodity to purposes of public surveillance does not exhaust the social-control implications of home networking. Chapter 4 has considered social control as social management. Unlike the privacy issue, which is an acknowledged social-concern, the social-control implications of social management have not crystallized as a public issue. Establishing the relationship between social management and home networking requires the integration of several lines of argumentation. Here the problem is not simply to define or identify practices of social management, but also to establish why these practices are undesirable. I have accomplished this by merging these two objectives in a historico-discursive analysis of the connection between computer systems and corporate power.

Stated more succinctly, the problem of defining social management and determining why it is objectionable is inextricably bound up with a critique of popular literature on computers and the question of how this issue is hidden within it. This critique is established at the levels of (1) systems analysis (the particular technocratic bias it gave to policy research, the myth of complexity); (2) the social theory of postindustrialism (how it validated systems analysis through its assumptions about the social order); and (3) futures research as the discourse which transposes the negative, manipulated connotations of social forecasting and planning into a benevolent project to keep the future open to human choice. The commentaries of Lewin and Michael were examined as a way of identifying social management as a particular orientation of social research, determined by the needs of the cybernetic economy of late capitalism and made possible in the postwar period by the data-processing and simulating capacities of the computer.

The critique of postindustrialism opened the way to a consideration of the macroeconomic environment which situates the introduction of home networking. Warskett's information economy of cybernetic competition, which represents a synthesis of a number of current schools of economic thought (particularly Mandel's work), reinforces my arguments concerning the social significance of home networking. Clearly, the informational commodity of home networking can be incorporated within and understood by macroeconomic theory as a potential input to the cybernetic economy. It is at this level that the value—and, by extension, the social significance—of home networking must be understood. The problem facing society is a serious one, because the inherent value of this information to the corporate sector may effectively foreclose debate on these systems before their political repercussions are fully understood. Indeed, I have already found evidence in the discourse on home networking that this has occurred in the form of a unidimensional conceptualization of social control which is overdetermined by the privacy issue.

The following scenario represents how the informational commodity will evolve as a component of a home networking industry. I assume that in the near future the push to develop an interactive services industry will come from hardware manufacturers, service providers, and communication carriers. The economic thrust will be to deliver services and finance the industry primarily through fees assessed to users. Initially, the informational commodity will be incidental to this financing structure. As home networking expands, the value of this information will be better understood; the use and resale of consumer information (whether in personal or aggregate form) will evolve as a component of the financing structure. Just how much value this information will add is a matter for speculation. Even when systems are in place, it may be difficult to calculate this in precise terms because, in many cases, the value of this information will depend on its input to other activities, such as marketing, and will be determined by their success.

The informational commodity will be applied to a variety of activities. It will have its most immediate applications in systems of public surveillance and in short-term social-management activities (i.e., marketing research oriented toward product planning, the creation of consumer profiles, consumer demographics, and interactive marketing). Behavior Scan suggests that interactive systems may permit marketers to integrate techniques of socio-demographic research with methods of behavioral psychology as the wired home becomes a virtual laboratory for market experimentation. These applications will have an immediate, tangible economic value for industry; they are already functional in a more costly form of survey research which must rely on telephone polling and census data. Eventually, this same information will be channeled into a qualitatively different order of social management practices oriented toward higher-level corporate policy making and planning. The focus of these activities is strategic planning for the long term and the engineering of consent for these initiatives. Interactive systems will provide more data on a more cost-effective basis as a way of improving long-term forecasting. Just as important, they will enable the corporate sector to monitor more efficiently public attitudes toward corporate policy as the repercussions of these initiatives unfold. To a great extent, such long-term forecasting may become a self-fulfilling prophecy as corporate resources are deployed to guarantee policy based on assumptions derived from long-term social and technological forecasting. The history of the videotex industry is itself a partial verification of this thesis. Technological optimism underscored numerous technological and market forecasts about the viability of interactive services in the late 1970s and early 1980s. Although these predictions were grossly divergent from economic and social reality, they succeeded in mobilizing sufficient amounts of capital and public-sector interest to launch the industry.

These practices of long-term social management represent a particular threat to a democratic society. It is one thing to manage the future allocation of material resources (although these activities are certainly a matter of public interest); it is quite a different thing, however, to engineer social response to corporate initiatives undertaken years before, to create and manipulate needs according to a corporate-sector agenda which has not been subject to public debate, and to systematically adjust the social order to the demands of corporate-sector hegemony. These activities of social management are vital to the functioning of an information economy of cybernetic competition. They are a matter of obvious public concern because they are based on a corporate-sector monopoly of knowledge which is, by definition, grounded in an inequality of knowledge. Attempts at public noncorporate intervention are frustrated, if not nullified, by an inability to command the knowledge resources implied by the planning economy.

North American society is clearly at a crossroads. Home networking is but one of a number of technological developments on the public agenda. The arguments presented here suggest, however, that home networking is a particulary important development, if not the major issue of the 1990s, given the repercussions of decisions made today for personal liberties, the balance of power between corporate and noncorporate sectors, and the manner in which our society will be governed in the future.

I have demonstrated that informed decision making on the future of home networking requires a level of social reflection which goes beyond a compartmentalized assessment of social impacts in isolated, unrelated contexts. Establishing the social significance of home networking as something latent in the current stage of its development is an important precondition to effective policy making. More research is called for, particularly that which integrates critical reflection with empirical research into the paths taken by personal information both within and among corporate bureaucracies. Efforts must also be made to educate the public and bring these issues to forums of public debate. These are matters of urgent public concern, but a more fundamental problem remains.

Home networking presents society with a set of radically different alternatives. Tracing the growth of the concept of home networking from Vannevar Bush's crude prototype of 1945 to the computer utility concept of the 1960s, one discovers a very early appreciation of the tremendous social potential of a system which would connect the home to an information and data processing network supported by the computer. Without equivocation, such a capacity was viewed as a social resource. Computer networks were discussed with reference to the public utility concept, the assumption being that a broad-based computer system should be at the service of society and its citizens. The model which informed the computer utility concept was based on experience with time-sharing computers which permitted remote access to a central computer using communication lines. This was an engineer's model to the extent that it was based primarily on the potential of technology. Economic, market, and political factors were not considered. After twenty years, home networking is becoming a reality, but it bears little resemblance to the system imagined by computer scientists of the 1960s.

Financing, promoting, and coordinating the development of home networking have proven difficult. A future home networking system would be a patchwork of technology, services, and competing industries, with little similarity to the neat model of the computer utility. As a result, it has become increasingly difficult to grasp home networking in its totality as a social resource of tremendous potential. In the current climate of deregulation and privatization, it may be considered utopian to hearken back to an older concept like the computer utility, but I believe that such an effort is justified. It is perhaps quixotic to imagine that one might reverse the engine of corporate-sector social research through prohibiting the use of aggregate social data. It is reasonable, however, to argue that this data be made available, within limits, to all of society—particularly because this information will emanate from a broad cross-section of it. A precondition for such a public information policy would be a political consensus on the status of information generated by home networking as a social resource. Given current trends toward the privatization of information, this option appears to be increasingly radical.

There is another radical alternative. Society may opt for the invisible hand of an imaginary market and let it determine communicational policy for home networking. Systems of public surveillance will be extended; corporate-sector social management will advance unchallenged; and the home will become the quintessential laboratory for social experimentation.

References
Index

References

Abt, Clark. 1970. An approach to methods of combined sociotechnological forecasting. *Technological Forecasting and Social Change* 2: 17–23.

Ackoff, Russell, and Patrick Rivett. 1963. *A manager's guide to operations research.* New York: Wiley.

Affluentials, a class market. 1983. *Marketing Communications,* Dec., 17–20.

Arlen, Gary. 1983a. Videotex battlefield. *Videotex Canada,* Winter, 15–19 ff.

Arlen, Gary. 1983b. Videotex goes on line. *Channels,* Nov./Dec., 40.

Ash, Steven, and John Quelch. 1982. *The new videotex technology and its impact on retailers in Canada.* Ottawa: Industry Trade and Commerce.

Balderston, E. F., J. M. Carman, and A. C. Hoggatt. 1977. Computers in banking and marketing. *Science,* 18 March, 1115–19.

Baran, Paul, and Paul Sweezy. 1966. *Monopoly capital.* New York: Monthly Review Press.

Baxter, William, Paul H. Cootner, and Kenneth E. Scott. 1977. *Retail banking in the electronic age: The law and economics of electronic funds transfer.* Monclair, N.J.: Allanheld, Osmun & Co.

Becker, Carl. 1945. *Freedom and responsibility in the American way of life.* New York: Knopf.

Bell, Daniel. 1973. *The coming of post-industrial society.* New York: Basic Books.

Berman, Paul. 1977. Computer or communications? In *High and low politics information resources for the 1980s,* ed. Anthony Oettinger et al., 147–94. Cambridge, Mass.: Ballinger.

Braverman, Harry. 1974. *Labor and monopoly capital.* New York: Monthly Review Press.

Brown, Ian. 1984. Videotex's uncertain fate. *Maclean's,* 14 May, 16b.

Brown, Michael, Brenda Billingsley, and Rebecca Shamai. 1980. *Privacy and personal data protection.* Research Publication 15. Ontario: Commission on Freedom of Information and Individual Privacy.

Brzezinski, Zbigniew. 1970. *Between two ages.* New York: Viking.

Buckley, Walter. 1969. *See* Lewin [1947] 1969.

Burns, Peter T. 1976. The law and privacy: The Canadian experience. *The Canadian Bar Review* 54(1): 1–12.

Bush, Vannevar. 1945. As we may think. *Atlantic Monthly,* July, 101–8.

Canada's top 10 multiple system operators. 1986. *Cable Communications Magazine* 52(11): 35.

Chaum, David. 1985. Security without identification. *Communications of the ACM* 28(1): 1032.

Clawson, Dan. 1980. *Bureaucracy and the labor process: The transformation of U.S. industry, 1860–1920.* New York: Monthly Review Press.

The computer age: Automating the boss' office. 1956. *Business Week,* 7 April, 59–62.

Cox, Michael. 1983. Personal access: The Canadian Human Rights Act of 1977 and the Privacy Act of 1982. In *Canada's new access laws: Public and personal access to government documents,* ed. Donald Rowat, 19–44. Ottawa: Department of Political Science, Carleton University.

Davitt, Lorena K. 1979–80. The Right to Financial Privacy Act: New protection for financial records. *Fordham Urban Law Journal* 8: 597–627.

De Sola Pool, Ithiel. 1983. *Technologies of freedom.* Cambridge, Mass.: Belknap Press.

De Sola Pool, Ithiel, and Robert Ableson. 1961. The simulatics project. *Public Opinion Quarterly* 25(2): 167–83.

Drucker, Peter. 1971. *The age of discontinuity.* New York: Pan Books.

Durivage, Paul. 1986. Vidéotron offrira un service d'achat. *Le Devoir,* 28 Aug., 11.

Ellis, James. 1986. Sears Discover Card starts to find its way. *Business Week,* 15 Sept., 166–67.

Etzioni, Amitar. 1968. *The active society.* New York: Free Press.

"Fiber/power cable" power. 1983. *Cable Communications,* July, 76.

Field, Anne, and Catherine Harris. 1986. The information business: Despite a slow start, many see a bonanza in selling electronic data. *Business Week,* 25 August, 82–90.

Flaherty, David. 1983. *Protecting privacy in two-way cable television services.* Toronto: Ontario Ministry of Transportation and Communications.

Garnham, Nicholas. 1986. The media and the public sphere. *Intermedia* 14(1): 28–33.

Gibson, Dale. 1980. Regulating the personal reporting industry. In *Aspects of privacy law,* ed. Dale Gibson, 111–40. Toronto: Butterworths.

Godfrey, David, et al., eds. 1979. *Gutenberg two.* Toronto: Press Porcepic.

Goldstein, Robert C., and Richard C. Nolan. 1975. Personal privacy versus the corporate computer. *Harvard Business Review* 53(2): 62–70.

Goldthorpe, John. 1971. Theories of industrial society: Reflections on the rescrudesence of historicism and the future of futurology. *Archives Europèenes de Sociologie 12(2): 263–88.*

Gotlieb, C. C. 1978. *Computers in the home.* Occasional Paper no. 4. Montreal: Institute for Research in Public Policy.

Greenberger, Martin. 1964. The computers of tomorrow. *Atlantic Monthly,* May, 63–67.

Greenberger, Martin, ed. 1968. *Computers and the world of the future.* Cambridge, Mass.: MIT Press.

Gross, Bertram. 1980. *Friendly fascism: The new face of power in America.* New York: M. Evans.

Habermas, Jurgen. 1970. Technology and science as "Ideology." In his *Toward a rational society: Student protest, science, and politics,* 81–122. Boston: Beacon Press.

Hallowell, John. 1971. *The decline of liberalism as an ideology.* New York: Howard Fertig.

Harnett, J. 1984. Transactional marketing. *Marketing Communications,* March, 36–41.

Harris, Marilyn. 1986. A lingua franca for computers. *Business Week,* 24 Feb., 110.

Herman, Andrew. 1981. *Corporate control, corporate power.* New York: Cambridge University Press.

Hondius, Frits. 1975. *Emerging data protection in Europe.* New York: American Elsevier.

Hoos, Ida. 1983. *Systems analysis in public policy: A critique.* Rev. ed. Berkeley and Los Angeles: University of California Press.

Innis, Harold. 1951. *The bias of communication.* Toronto: University of Toronto Press.

Ivey, Mark, M. Pitzer, K. Dreyfack, and M. Vamos. 1986. Home shopping is it a revolution in retailing—or just a fad? *Business Week,* 15 Dec., 62–69.

Jensen, Jay. 1976. *Liberalism, democracy and the mass media.* Urbana-Champaign: Univ. of Illinois Press.

Kahn, Herman, and A. Weiner. 1967. *The year 2000.* New York: Macmillan.

Keller, John, P. Engardio, K. Dreyfack, and R. Mitchell. 1986. The rewiring of America: Making phone lines the transportation system for the information age. *Business Week,* 15 Sept., 188–96.

Kerr, Elaine, and Starr Roxanne Hiltz. 1982. *Computer-mediated communication systems.* New York: Academic.

Koughan, Martin. 1981–82. The state of the revolution, 1982. *Channels,* Dec./Jan., 23–29.

Kumar, Krishan. 1978. *Prophesy and progress: The sociology of industrial and post-industrial society.* London: Penguin.

Lane, Michael. 1984. Public access videotex. *Videotex Canada,* Spring, 17–21.

Lesage, Gilles. 1983. Accès aux documents publics. *Le Devoir,* 19 Nov., 1.

Lewin, Kurt. [1947] 1969. Feedback problems of social diagnosis and action. In *Modern systems research for the behavioral scientist,* ed. Walter Buckley, 441–44. Chicago: Aldine.

Linstone, Harold, and Ralph C. Lenz. 1970. The second year. *Technological Forecasting and Social Change* 2:1.

MacBride, Robert O. 1967. *The automated state: Computer systems as a new force in society.* Philadelphia: Chilton Book Co.

Madden, John. 1979. *Videotex in Canada.* Ottawa: Minister of Supply and Services.

Magliana, John A. 1984. The smart card. *Marketing Communications,* May, 48–50.

Mandel, Ernest. 1978. *Late capitalism.* London: Verso.

Marcuse, Herbert. 1969. Industrialization and capitalism in the work of Max Weber. In his *Negations, essays in critical theory,* 201–26. Boston: Beacon Press.

Marglin, Stephen A. 1974. What do bosses do? *Review of Radical Political Economics* 6(2): 33–60.

McCarthy, John. 1968. Time-sharing computer systems. In *Computers and the world of the future,* ed. Martin Greenberger, 220–48. Cambridge, Mass.: MIT Press.

Michael, Donald. 1967. Social engineering and the future environment. *American Psychologist* 40:888–92.

Mickelwright, Christine. 1982. Major issues for the 1980's—health, technology, and employment. A brief submitted to the Labour Canada Task Force on Micro-technology and Employment by the Brotherhood of Railway and Airline Clerks.

Miles, Ian. 1978. The ideologies of futurists. In *Handbook of futures research,* ed. Jib Fowles, 5–15. Westport, Conn.: Greenwood Press.

Miliband, Ralph. 1969. *The state in capitalist society.* New York: Basic Books.

Moorecraft, Kip. 1986. Cable home shopping: Impulse-buy TV that works. *Cable Communications Magazine* 52(9): 22–28.

Morse, R. C. 1982. The home information explosion. Parts 1, 2. *Marketing Communications,* Feb., 24–30; March, 24–29.

Mosco, Vincent. 1982. *Pushbutton fantasies: Critical perspectives on videotex and information technology.* Norwood, N.J.: Ablex.

Mowshowitz, Abbe. 1976. *The conquest of will: Information processing in human affairs.* Don Mills, Ontario: Addison-Wesley.

Mowshowitz, Abbe. 1983. Scholarship and policy making: The case of computer communications. *Computers and Society* 13 (Spring): 2–9.

Nahon, George. 1986. Videotex in France: Status report as of March 1986. *Intelmatique France,* March, 1–6.

Nash, Deanna Collingwood, and John B. Smith. 1981. *Interactive home media and privacy.* Washington, D.C.: Federal Trade Commission, Office for Planning.

A new course at Sears. 1983. *Marketing Communications,* Oct., 17–21.

Noble, David. 1977. *America by design: Science, technology, and the rise of corporate capitalism.* New York: Knopf.

Now merchants can learn more about charge customers. 1984. *Business Week,* 19 March, 47.

O'Toole, John. 1981. TV commercials: The new frontier. Advertisers can take advantage of modern video technology by replacing mass messages with targeted appeals. *Marketing Communications,* Dec., 31–34.

On-Line Databases. 1983. *Videotex Canada,* Winter, B1–B4; D1–D2; G2–G7.

Parkhill, Douglas. 1979. The necessary structure. In *Gutenberg two,* ed. David Godfrey and Douglas Parkhill, 69–95. Toronto: Press Porcepic.

Pergler, P. 1980. *The automated citizen.* Occasional Paper no. 14. Montreal: The Institute for Research in Public Policy.

Privacy Protection Study Commission. U.S. Congress. 1977. *Personal privacy in an information society.* Washington, D.C.: GPO.

Provost, Gilles. 1984. Le groupe Vidéotron a déjà commandé 100,000 décodeurs chez Vidacom." *Le Devoir,* 24 Jan., 9.

Rubin, Ken. 1983. *Prying eyes: A review and indictment of the 1982 Canadian Privacy Act.* Ottawa: Ken Rubin.

Rule, James. 1973. *Private lives and public surveillance: Social control in the computer age.* London: Allen Lane.

Schiller, Herbert. 1981. *Who knows: Information in the age of the Fortune 500.* Norwood, N.J.: Ablex.

Schmidt, Joseph S., and Ruth B. Corbin. 1983. Telecommunications in Canada: The regulatory crisis. *Telecommunications Policy,* Sept., 215–27.

Schmitter, Philippe. 1979. Still the century of corporatism. In *Trends towards corporatist intermediation,* ed. Phillipe Schmitter and G. Lehmbruch, 7–51. Beverly Hills: Sage.

Science Council of Canada. 1979. *A scenario for the implementation of interactive computer communications systems in the home.* A Position Paper of the Science Council Committee on Communications and Computers, July. Ottawa: Science Council.

Shifting sales into high gear. 1984. *Marketing Communications,* Jan., 15–21.
Sigel, Efrem, ed. 1980. *Videotext: The revolution in home/office information retrieval.* White Plains, N.Y.: Knowledge Industry Publications.
Siwolop, Sana. 1987. Excuse me, what's the pollsters' big problem? *Business Week,* 16 Feb., 108.
Smith, Anthony. 1980. *Goodbye Gutenberg.* New York: Oxford University Press.
Stanbury, W. T., and Fred Thompson. 1982. *Regulatory reform in Canada.* Montreal: Institute for Research in Public Policy.
Task Force on Privacy and Computers. Canada, Department of Justice and Department of Communications. 1973. *Privacy and Computers.* Ottawa: Information Canada.
Telidon content development grants. 1984. *Videotex Canada,* Spring, 48.
Trudel, Clément. 1983. Il y a risque de conflit entre l'Etat et la commission d'acces à l'information. *Le Devoir,* 19 Nov., 12.
Tydeman, John, H. Lipinski, R. Adler, M. Nyhan, and L. Zwimpfer. 1982. *Teletext and videotex in the United States: Market potential, technology, policy issues.* New York: McGraw-Hill.
U.S. Department of Health, Education, and Welfare. 1973. Secretary's Advisory Committee on Automated Personal Data Systems. *Records, computers and the rights of citizens.* Washington, D.C.: GPO.
Vista—what's over the horizon? 1983. *Videotex Canada,* Winter, 26–29.
Walkom, Thomas. 1984. Privacy act "Endangered" by Provinces. *Globe and Mail,* 5 July, 1.
Warskett, George. 1981. The information economy in late capitalism. In *Communication studies in Canada,* ed. Liora Salter, 178–95. Toronto: Butterworths.
Weber, Max. 1946. *From Max Weber: Essays in sociology.* Ed. and trans. H. H. Girth and C. Wright Mills. New York: Oxford University Press.
Weinstein, A. 1976. The privacy act of 1974: An overview and critique. *Washington University Law Quarterly,* Fall, 667–718.
Weizenbaum, Joseph. 1976. *Computer power and human reason.* San Francisco: Freeman.
Weissman, Steven. 1983. Videotex: A technology in search of a market. *Marketing Communications,* Sept., 30–36.
Westin, Alan. 1967. *Privacy and freedom.* New York: Atheneum.
Westin, Alan. 1982. Home Information Systems: The Privacy Debate. *Datamation* 28(1): 100–114.
Westin, Alan, and Michael Baker. 1972. *Databanks in a free society: Computers, record-keeping, and privacy.* National Academy of Sciences, Washington, D.C., Project on Computer Databanks. New York: Quadrangle Books.
Williams, Raymond. 1961. *The long revolution.* London: Penguin.
Williams, Raymond. 1974. *Television technology and cultural form.* Glasgow: Fontana/Collins.
Wilson, Kevin G. 1982. Futures research and interactive videotex: An analysis of the discourse on new communications technology and society. Unpublished paper.
Window on the world: The home information explosion. 1981. *Business Week,* 29 June, 74–83.
Wynne, B. 1975. The rhetoric of consensus politics. *Research Policy* 4: 109–58.

Additional Sources

Arrow, Kenneth. 1979. The economics of information. In the *The computer age,* ed. Michael Dertouzos and J. Moses, 306–17. Cambridge, Mass.: MIT Press.

Auerbach, Lewis. 1983. Privacy and Canadian telecommunications regulation. *Telecommunications Policy, March, 39–40.*

Bamford, James. 1983. *The puzzle palace.* New York: Penguin.

Beer, Stafford. 1967. *Decision and control.* New York: Wiley.

Bell, Daniel. 1977. Teletext and technology. *Encounter,* June, 9–29.

Bellman, R. 1970. Man, society, and the computer. In *Man and computer,* Proceedings of the First International Conference on Man and Computer, ed. M. Marois, 229–45. 22–26 June, Bordeaux. Paris: Unesco.

Blau, Peter. 1955. *The dynamics of bureaucracy.* Chicago: University of Chicago Press.

Block, F., and L. Hirschorn. 1979. New productive forces and the conditions of contemporary capitalism: A post-industrial perspective. *Theory and Society* 7: 363–95.

Bloustein, Edward J. 1977. Group privacy: The right to huddle. *Rutgers-Camden Law Journal* 8: 219–83.

Boguslaw, Robert. 1965. *The new utopians: A study of system design and social change.* Englewood Cliffs, N.J.: Prentice-Hall.

Boorman, Scot. 1983. Block models and self-defence: The computer as judge and jury. *New York Times,* 27 Nov., 3.

Boulden, James B., and Elwood Buffa. 1970. Corporate computer: Online, realtime systems. *Harvard Business Review,* 48(4): 65–83.

Bright, James. 1958. *Automation and management.* Boston: Harvard Business School.

Brock, Gerald. 1975. *The U.S. computer industry: A study of market power.* Cambridge, Mass.: Ballinger.

Brock, Gerald. 1981. *The telecommunications industry: The dynamics of market structure.* Cambridge, Mass.: Harvard University Press.

Burck, Gilbert. 1964. The boundless age of the computer. *Fortune,* March, 101–10.

Burck, Gilbert. 1964. "On Line" in "Real Time." *Fortune,* April, 141–45.

Burck, Gilbert. 1964. Management will never be the same again. *Fortune,* Aug., 125–30.

Burns, Peter T. 1980. Privacy and common law: A tangled skein unraveling? In *Aspects of privacy law,* ed. Dale Gibson, 21–40. Toronto: Butterworths.

Cable's top ten. 1983. *Channels,* Nov./Dec., 26.

Canada. Department of Communications. 1973. *Survey of public attitudes towards the computer.* Ottawa.

Chevreau, Jonathan. 1983. Most TV networks stay with experiments. *Globe and Mail,* 21 Oct., B14.

Churchman, C. West. 1979. *The systems approach.* Rev. ed. New York: Dell.

Clegg, Stewart. 1975. *Power, rule and domination: A critical and empirical understanding of power in sociological theory and organizational life.* London: Routledge & Kegan Paul.

Crozier, Michel. 1964. *The bureaucratic phenomenon.* Chicago: University of Chicago Press.

Deutsch, Karl. 1968. Toward a cybernetic model of man and society. In *Modern systems research for the behavioral scientist,* ed. Walter Buckley, 387–400. Chicago: Aldine.

Duncan, Otis Dudley. 1969. Social forecasting—the state of the art. *Public Interest* 17: 88–118.

Edwards, Richard C. 1975. Stages in corporate stability and the risks of corporate failure. *Journal of Economic History* 35(2): 248–57.

Edwards, Robert. 1979. *Contested terrain: The transformation of the workplace in the twentieth century.* New York: Basic Books.

The exploding home computer population and the future of two-way cable. 1983. *Cable Communications,* Sept., 26–28.

Fenichell, Stephen. 1982. The push for electronic security. *Channels,* June/July, 50–52.

Fishman, Katherine. 1981. *The computer establishment.* New York: McGraw-Hill.

Flaherty, David. 1979. *Privacy and government data banks. An international perspective.* London: Mansell.

Frees, Jan. 1982. The Swedish data act. *Current Sweden,* Sept.: 2–9.

Friedrichs, Guenter, and Adam Schaff, eds. 1982. *Micro-electronics and society: A report to the Club of Rome.* Scarborough, Ont.: Mentor.

Galbraith, John. 1978. *The new industrial state.* 3rd ed. New York: New American Library.

Ganley, Oswald. 1979. Communications and information resources in Canada. *Telecommunications Policy,* Dec., 267–88.

Gilchrist, Bruce, and Richard Weber. 1973. *The state of the computer industry in the United States.* Montvale, N.J.: American Federation of Information Processing Societies Inc.

Gitlin, Todd. 1981. New video technology: Pluralism or banality. *Democracy* 1(4): 60–76.

Glen, H. Patrick. 1980. The right to privacy in Quebec law. In *Aspects of privacy law,* ed. Dale Gibson, 41–72. Toronto: Butterworths.

The golden plan of American Express. 1984. *Business Week,* 30 April, 118–22.

Goldstine, Herman. 1972. *The computer from Pascal to Von Neuman.* Princeton: Princeton University Press.

Grace, John W. 1983. Public laws for private lives. Notes for an address to the Conference on Human Rights from a Canadian perspective. Acadia University, Nova Scotia, 4 Oct.

Grant, George. 1976. The computer does not impose on us the ways it should be used. In *Beyond industrial growth,* ed. Abraham Rotstein, 117–31. Toronto: University of Toronto Press.

The growth potential of "Supertex." 1983. *Cable Communications,* April, 56–57.

Habermas, Jurgen. 1975. Crisis tendencies in advanced capitalism. In his *Legitimation crisis,* 33–92. Boston: Beacon.

Hansen, Inger. 1983. Notes for an address on the role of the information commissioner. Toronto, Ontario, 15 Sept.

Harris, William. 1957. The electronics business. *Fortune,* April, 137–43.

Haslam, Gerald. 1980. Information provider activities in Canada. Ottawa: Dept. of Communications, Aug.

Herman, Andrew. 1982. Conceptualizing control: Domination and hegemony in the capitalist labor process. *The Insurgent Socialist* 11(3): 7–22.

Hoffman, Lance, ed. 1980. *Computers and privacy in the next decade.* New York: Academic Press.

Holsendolph, Ernest. 1983. Senate panel clears compromise bill on cable TV. *New York Times,* 22 April, B5.

Hoos, Ida. 1960. When the computer takes over the office. *Harvard Business Review,* July–Aug., 102–12.

Instant world: A report on telecommunications in Canada. 1971. Ottawa: Information Canada.

Irwin, Manley R., and Steven Johnson. 1977. The information economy and public policy. *Science* 195(4283): 1170–74.

Janowitz, Morris. 1978. The intellectual history of "social control." In *Social control for the 1980s,* ed. J. Roucek, 20–45. Westport, Conn.: Greenwood Press.

Jessop, Bob. 1979. Corporatism, parliamentarism, social democracy. In *Trends toward corporatist intermediation,* ed. P. C. Schmitter and G. Lehmbruch, 185–211. London: Sage.

Jessup, Bob. 1982. *The capitalist state.* New York: New York University Press.

Katz, Frede. 1968. *Autonomy and organization: The limits of social control.* New York: Random House.

Kay, Peg. 1978. Policy issues of interactive cable television. *Journal of Communications* 28: 202–8.

Krippendorff, Klaus, ed. 1979. *Communication and control in society.* New York: Gordon & Breach.

Lasch, Christopher. 1972. Toward a theory of post-industrial society. In *Politics in the post-welfare state,* ed. M. D. Hancock and G. Sjoberg, 36–50. New York: Columbia University Press.

Leiss, William. 1974. *The domination of nature.* Boston: Beacon.

Lemelshtrich, Noam. 1974. *Two-way communication: Political and design analysis of a home terminal.* London: Sage.

Lesser, Barry. 1982. *Alternative market structure for videotex service in Canada. The public policy implications.* Ottawa: Department of Communications.

Marris, Robin. 1964. *The economic theory of "managerial" capitalism.* Glencoe, Ill.: The Free Press.

Martindale, Don. 1978. The theory of social control. In *Social control for the 1980s,* ed. J. Roucek, 46–58. Westport, Conn.: Greenwood Press.

McCarthy, J. 1970. The home information terminal. In *Man and computer,* Proceedings of the First International Conference on Man and Computer, ed. M. Marois, 48–57. 22–26 June, Bordeaux. Paris: Unesco.

McNiven, J. D. 1976. Bureaucratic imperatives and citizen access: Some theoretical possibilities. *Optimum* 7(3): 22–32.

Mersich, Dan. 1981. Will telematics really change our world? *Canadian Data Systems,* June, 32.

Michael, Donald. 1964. Speculations on the relation of the computer to individual freedom and the right to privacy. *George Washington Law Review* 33 (Oct.): 270–86.

Michael, Donald. 1966. Some long-range implications of computer technology for human behavior in organizations. *American Journal of Behavioral Science* 9 (April): 29–35.

Michael, Donald. 1968. On coping with complexity: Planning and politics. *Daedalus* 97(Fall): 1179–93.

Miller, Arthur R. 1971. *The assault on privacy.* Ann Arbor, Mich.: University of Michigan Press.

Miller, Arthur S. 1973. Privacy in the modern corporate state: A speculative essay. *Administrative Law Review* 25 (Summer): 231–67.

Miller, Arthur S. 1976. *The modern corporate state.* Westport, Conn.: Greenwood Press.

Miller, S. M. 1975. Notes on neo-capitalism. *Theory and Society* 2(1): 1–35.

Mizruchi, Mark S. 1982. *The American corporate network, 1904–74.* Beverly Hills: Sage.

Morse, Philip M., ed. 1967. *Operations research for public systems.* Cambridge, Mass.: MIT Press.

Mouzelis, Nicos P. 1967. *Organization and bureaucracy: An analysis of modern theories.* London: Routledge and Kegan Paul.

Mowshowitz, Abbe. 1980. Ethics and cultural integration in a computerized world. In *Human choice and computers, 2,* ed. A. Mowshowitz, 251–69. Amsterdam: North-Holland.

Myers, Charles A. 1967. *The impact of computers on management.* Cambridge, Mass.: MIT Press.

New media: Broadcast teletext, videotex. 1981. Summary of the final report presented by the commission on New Information Technology. Stockholm, Sweden.

New views: Computers and new media—anxiety and hopes. 1979. Report by the Commission on New Information Technology. Stockholm, Sweden.

Nora, Simon, and Alain Minc. 1981. *The computerization of society.* Cambridge, Mass.: MIT Press.

Organisation for Economic Co-operation and Development (O.E.C.D.). 1977. *Transborder data flows and the protection of privacy.* Proceedings of a symposium held in Vienna, 20–23 Sept. Paris: O.E.C.D.

Oettinger, Anthony, et al. 1977. *High and low politics: Information resources for the 80's.* Cambridge, Mass.: Ballinger.

Pecujlic, M. 1970. Cybernetics—Social Power—Man. In *Man and computer,* Proceedings of the First International Conference on Man and Computer, ed. M. Marois, 200–219. 22–26 June, Bordeaux. Paris: Unesco.

Players: Powers that be. 1983. *Channels, Field Guide 1984,* Nov./Dec., 64–65.

Pollan, Michael. 1982. The electronic library. *Channels,* June/July, 30–31

Porat, Marc U. 1978. Communication policy in an information society. In *Communications for tomorrow: Policy perspectives for the 1980s,* ed. Glen O. Robinson, 3–60. New York: Praeger.

Poulantzas, Nicos. 1973. *Political power and social classes.* London: New Left Books.

Provost, Gilles. 1984. La presse bénéficiera d'un nouveau programme Télidon. *Le Devoir,* 13 Jan., 3.

Rapport du Commissaire à la protection de la vie privée sur l'utilisation du numéro d'assurance sociale. 1981 (Jan.). Canadian Human Rights Commission, Privacy Commissioner.

Redmond, Kent C., and Thomas Smith. 1980. *Project Whirlwind: The history of a pioneer computer.* Bedford, Mass.: Digital Press.

Report of the Commission of Inquiry into the confidentiality of health information. 1980. 3 vols. Toronto: Queen's Printer.

Rieser, Carl. 1962. The short-order economy. *Fortune*, Aug., 90–94.

Roucek, Joseph. 1978. The concept of social control in American Sociology. In *Social control for the 1980s*, ed. J. Roucek, 3–19. Westport, Conn.: Greenwood Press.

Rowat, Donald, ed. 1980. *The right to know: Essays on governmental publicity and public access to information.* Ottawa: Carleton University, Department of Political Science.

Rule, James, Douglas McAdam, Linda Stearns, and David Uglow. 1980. *The politics of privacy.* New York: New American Library.

Scase, R., ed. 1977. *Industrial society: Class, cleavage, and control.* London: Allen & Unwin.

Schmitter, Philippe C. 1979. Still the century of corporatism? In *Trends toward corporatist intermediation*, ed. P. C. Schmitter and G. Lehmbruch, 7–52. London: Sage.

Schroyer, Trent. 1973. *The critique of domination: The origins and development of critical theory.* New York: George Braziller.

Science Council of Canada. 1982. *Planning now for an information society: Tomorrow is too late.* Hull, Quebec: Minister of Supply and Services.

Serafini, Shirley, and Michel Andrieu. 1981. *The information society and its implications for Canada.* Hull, Quebec: Minister of Supply and Services.

Simon, Herbert. 1960. *The new science of management decision.* New York: Harper and Row.

Simon, Herbert. 1979. The consequences of computers for centralization and decentralization. In *The computer age*, ed. Michael Dertouzos and Joel Moses, 212–28. Cambridge, Mass.: MIT Press.

Smith, Robert Ellis. 1979. *Privacy: How to protect what's left of it.* Garden City, N.Y.: Anchor/Doubleday.

Stern, Nancy. 1981. *From Eniac to Univac: An appraisal of the Eckert-Mauchly computers.* Bedford, Mass.: Digital.

Stothers, Jennifer. 1983. The prospects for home shopping services on videotex. *Videotex Canada*, Fall, 8–11.

Strachey, C. 1959. Time sharing in large fast computers. *Proceedings of the International Conference on Information Processing*, ed. S. De Picciotto, 336–41. Held June, 15–20 Paris. Paris: Unesco.

Swerdlow, Joel. 1981. Why is everybody afraid of Ma Bell? *Channels*, Oct./Nov., 29–33.

Taylor, Frederick W. 1947. *Scientific management.* New York: Harper and Row.

Three big backers give videotex a shot at success. 1984. *Business Week*, 27 Feb., 36.

Umpleby, Stuart A. 1970. Citizen sampling simulations: A method for involving the public in social planning. *Policy Sciences* 1: 361–75.

Umpleby, Stuart A. 1972. Is greater citizen participation in planning possible and desirable? *Technological Forecasting and Social Change* 4: 61–76.

Van Deerlin, Lionel. 1982. How Ma Bell choreographed the settlement. *Channels*, April/May, 41–43.

Videotex hits the road. 1983. *Telecommunications*, Sept., 116.

Vitalis, André. 1981. *Informatique, pourvoir et libertés.* Paris: Economica.

Weber, Max. 1947. *The theory of social and economic organization.* Trans. A.M. Henderson and Talcott Parsons. London: Free Press of Glencoe.

Wells, Diane, and Chris Dornan. 1981. Videotex: Availability to users. *Working Papers in Communications.* Montreal: McGill University Graduate Program in Communications.

Wells, Lucy. 1983. *Les implications administratives des législations étrangères en matière d'accèss à l'information et de protection de la vie privée: le cas de la France, de la Suède, des Etats Unis et du Canada en perspective avec celui du Québec.* Québec: Ministère des Communications.

Wessel, Andrew. 1976. *The social use of information: Ownership and access.* New York: Wiley.

Whisler, Thomas. 1970. *The impact of computers on organizations.* New York: Praeger.

Wicklein, John. 1979. *The electronic nightmare: The new communications and freedom.* New York: Viking.

Woolfe, Roger. 1980. *Videotex: The new television/telephone information services.* London: Hyden.

Zimbalist, Andrew, ed. 1979. *Case studies in the labor process.* New York: Monthly Review Press.

Index

COMPOSED BY CONNELL TYPESETTING COMPANY, KANSAS CITY, MISSOURI
MANUFACTURED BY EDWARDS BROTHERS, INC., ANN ARBOR, MICHIGAN
TEXT AND DISPLAY LINES ARE SET IN TIMES ROMAN

Library of Congress Cataloging-in-Publication Data
Wilson, Kevin G., 1952–
Technologies of control.
(Studies in communication and society)
Bibliography: pp. 161–171.
Includes index.
1. Videotex systems—Social aspects—United States.
2. Interactive computer systems—Social aspects—United States. 3. Privacy, Right of—United States. I. Title.
II. Series.
TK5105.W56 1988 303.4′834′0973 87-40378
ISBN 0-299-11370-1
ISBN 0-299-11374-4 (pbk.)